THE LIGHT OF THE NIGHT SKY

GEOPHYSICS AND ASTROPHYSICS MONOGRAPHS

AN INTERNATIONAL SERIES OF FUNDAMENTAL TEXTBOOKS

VOLUME 4

THE LIGHT
OF THE NIGHT SKY

by

F. E. ROACH

Rutgers, The State University, Newark, N.J., U.S.A.;
University of Hawaii, Honolulu, H.I., U.S.A.

and

JANET L. GORDON

Bernice P. Bishop Museum, Honolulu, H.I., U.S.A.

D. REIDEL PUBLISHING COMPANY

DORDRECHT-HOLLAND / BOSTON-U.S.A.

First printing: December 1973

Library of Congress Catalog Card Number 73–83568

Cloth edition: ISBN 90 277 0293 4
Paperback edition: ISBN 90 277 0294 2

Published by D. Reidel Publishing Company,
P.O. Box 17, Dordrecht, Holland

Sold and distributed in the U.S.A., Canada, and Mexico
by D. Reidel Publishing Company, Inc.
306 Dartmouth Street, Boston,
Mass. 02116, U.S.A.

Printed in The Netherlands by D. Reidel, Dordrecht

(Physics)

TABLE OF CONTENTS

TABLES

PLATES

ILLUSTRATIONS

Zodiacal light seen from Haleakala, Hawaii, January 1967.

(Photograph by Courtesy of P. B. Hutchison, Dudley Observatory.)

PREFACE

Astronomy appears to us as a combination of art, science, and philosophy. Its study puts the universe into perspective, giving a sense of pleasure in its beauty, awe at its immensity, and humility at our trivial place in it.

From earliest human history, man has scrutinized the night sky – and wondered and marveled. With unaided eye but perceptive mind, he recognized order in the regular appearance and movements of individual objects, such as the planets and star groups (constellations), in their rhythmic and majestic progressions across the bowl of night. Even in the present era of scientific exactitude, there remains a profound awareness of mysteries beyond our present interpretations.

It is only in comparatively recent years, however, that man has recognized that it takes more than conventional astronomy to account for the beauties of the night sky. Radiations in the Earth's upper atmosphere provide a foreground light, the study of which has come under a new name, aeronomy. The science of aeronomy has rapidly burgeoned, and the student of the light of the night sky finds that he is involved in an interdisciplinary domain.

The 'meat' of one discipline, however, may be the 'poison' of the other. To the astronomer, the Earth's atmosphere, inhibiting his extra-terrestrial viewing, is a serious nuisance. To the aeronomer, the Moon, planets, stars, and Galaxies hamper his measurements and interfere with his studies of the Earth's upper atmosphere. Yet both sets of elements are basic to the beauties as well as to the understanding of the light of the night sky.

It is essentially the students of astronomy and aeronomy for whom we have written this book. We also hope, however, that it will present much of interest and value to the bemused sky watcher, for whom some detailed knowledge of the several con-tried to meld these dual objectives to create a broadly based, professionally valid tributors to the nighttime sky may increase his pleasure in contemplating it. We have treatise that will lead the serious student to deeper probing into the phenomena and will inspire both him and the enthusiastic amateur to an appreciation of that half of their experience which we may refer to as their 'night life'.

As any scientific book goes to press, it becomes obsolete – knowledge continues to expand along with the expanding universe. Superimposed on the obsolescence inherent in an active subject being reported at a point in time is another; the night sky cannot be considered constant in time scales – astronomical, geological, or even historical. Our view toward any part of the Galaxy changes for two reasons – first, the differential rotation of layers outward from the center and, second, the fact that the intrinsically very bright stars have lifetimes relatively short, measured in only millions of years. Thus during a period of one galactic rotation of 200 000 000 yr (at our distance from the galactic center) – less than 100 times man's existence on Earth – the appearance of the Milky Way must have changed dramatically.

In Chapter 6 we call attention to the fact that the zodiacal cloud, which is responsible for an important component of the light of the night sky, has a lifetime in the order of 80 000 years. Astronomically speaking, this is a short-lived phenomenon.

A factor somewhat under human control is the radiation from the upper atmosphere. Already it has been measurably affected several times by changes in the ionization of the upper atmosphere from high-altitude nuclear detonations. Lower-atmosphere contaminations due to the 'civilization' crawling over the planet's surface could conceivably affect our viewing of the night sky in the centuries ahead.

Our description of the several components contributing to the light of the night sky applies to the present time and to our particular cosmic location. We thus present a study of an ephemeral and parochial phenomenon.

Honolulu, August 1973

CHAPTER 1

FROM DAY TO TWILIGHT TO NIGHT

Life on planet Earth is tuned to the diurnal drama of day-twilight-night-dawn-day. As Earth rotates within the sunlight sheath, the outward-looking observer notes changes of brightness which cover many powers of ten. The day sky, in the absence of clouds, appears as a blue vault resulting from the scattering of sunlight by the atmospheric components. The brightness of the sky decreases rapidly after sunset as the solar rays illuminate only the higher regions of the atmosphere above the observer. At a solar depression of 18° (end of astronomical twilight) the amount of illuminated atmosphere has become so small that the scattered sunlight is trivial and the night sky can be observed without twilight competition (Figure 1-1).

1.1. The Day Sky – The Blue Planet

The blueness of the day sky was explained by Lord Rayleigh III in his classical paper, 'On the Light from the Sky, Its Polarisation and Colour' (1871), the key portion of which is presented in Plate I. Analytically, he found that, for small particles,

$$\frac{I}{I_0} = \frac{\text{const.}\,(1+\cos^2\theta)}{\lambda^4}. \tag{1.1}$$

Since the human eye is sensitive over the octave from about 7000 Å $(7 \times 10^{-5}$ cm$)$ to 3500 Å $(3.5 \times 10^{-5}$ cm$)$, the ratio of intensity of scattered light for the shorter (bluer) wavelength relative to the longer (redder) wavelength is

$$\frac{I_{\text{blue}}}{I_{\text{red}}} = \left(\frac{\lambda_{\text{red}}}{\lambda_{\text{blue}}}\right)^4 = \left(\frac{7000}{3500}\right)^4 = \left(\frac{2}{1}\right)^4 = 16. \tag{1.2}$$

In Equation (1.1) the term $(1+\cos^2\theta)$ is due to two conditions of polarization – the 1 corresponds to the electric vector perpendicular to the plane of scattering, and the $\cos^2\theta$ to the electric vector parallel. The situation is best shown in the form of a polar diagram (Figure 1-2). Most of the incident radiation will pass the particle in its path, but a small fraction will be scattered. The relative intensity of the total scattered light as a function of scattering angle is shown as the solid line $(1+\cos^2\theta)$. The two polarized components are shown as dashed lines. Polar diagrams of scattered light from particles comparable in size to the wavelength (such as dust) will be discussed later in connection with two astronomical phenomena – the zodiacal light and the diffuse galactic light.

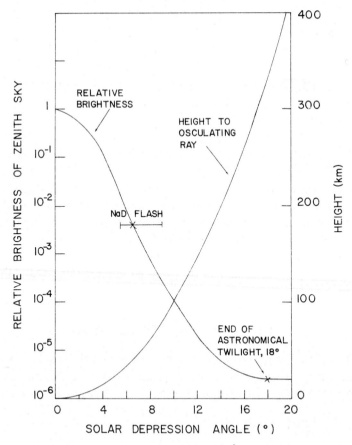

Fig. 1-1.   Relative brightness of zenith sky during twilight (5200 Å; after Ashburn, 1952:85), and height
to lowest part of zenith atmosphere illuminated by Sun during twilight.

1.2. The Twilight Sky

As Earth's rotation carries the Sun from the horizon to an 18° depression angle, the
zenith brightness decreases about 400 000-fold, as shown in Figure 1-1. In the tropics,
this darkening can occur in as short a time as 72 min.

   With the increasing depression angle, progressively higher levels of the upper atmosphere are being irradiated by the Sun, until the night level of 'brightness' is achieved.
The general spectroscopic nature of the light during the twilight phase is that of
scattered sunlight. A color photograph during midtwilight reveals a blue sky long
after the eye has lost its color sensitivity, showing that the mechanism is still Rayleigh
scattering.

   At a solar depression of about 6° the lower boundary of the solar radiation is at
35-km altitude. Solar radiation impinging on the sodium atoms above this region ex-
cites them to the $^2P^0$ state, which results in the emission of the yellow Na D lines. At a
solar depression of 6.5°, the general twilight radiation has been reduced to such a low

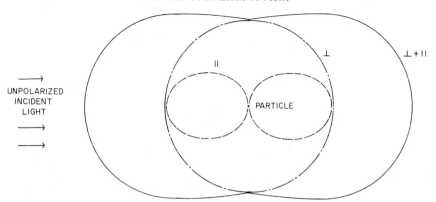

Fig. 1-2.   Polar diagram of scattering of unpolarized incident light by small particle (e.g., molecule). The
components with electric vector perpendicular ($\perp$) and parallel ($\parallel$) to the scattering plane, and
their sum are shown.

photometric level that the emission lines are of intensity equal to that of the absorption Na D lines in the solar spectrum (Figure 1-1). The lines fade in brightness rapidly as the increasing solar-depression angle puts the osculating ray above the level at which the Na atoms are sufficiently numerous to produce an observable line. The relatively short duration of the phenomenon has led to its description as the 'sodium flash'. A similar flash occurs during twilight for the red (6300 Å, 6364 Å) lines of O.

1.3. The Human Eye and Its Adaptation to Low Light Levels

The eye, with its remarkable ability to adapt to the very wide range of brightness from the day to the night sky levels, adds a dimension of aesthetic enjoyment to the 'sky watcher'.

The dark adaptation of the eye is a complicated process involving a small factor (of about seven) that is due to the opening up of the iris to admit more light, and a very large factor (which can be more than 10^3, for a large field) that results from a chemical change in one of the two types of light receptors in the retina. These receptors – named rods and cones because of their shapes – serve different functions.

The approximately 7×10^6 cones are concentrated in the central part of the retina, directly behind the lens, and provide very sharp visual definition and color sensitivity over a small field – 2 to 5° – at high light levels. They decrease in frequency outward from that center.

The 13×10^7 rods have the complementary distribution, being sparse near the center of the retina but increasing in numbers and concentrating as they approach the periphery of the retina. They generate the unique chemical 'visual purple', which they accumulate in the absence of light and which seems to be the active agent in achieving dark-adapted vision over a period of time. The rods detect faint light and movement more effectively than do cones. Thus, in dark adaptation, they are the active element.

Figure 1-3 shows the relative sensitivity of the eye to dark adaptation as a function of time for five centrally fixated fields varying from 2 to 20°. It is striking to realize that,

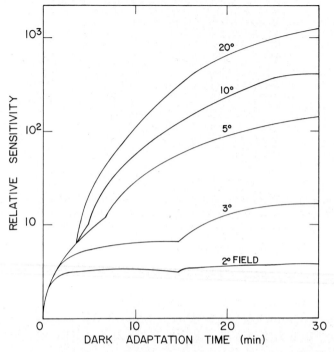

Fig. 1-3.   Increase in visual sensitivity with dark-adaptation time for a centrally fixated field
of variable diameter.

over the wider fields – from 5 to 20° – the relative sensitivity increases between 10^2 and 10^3 times in a period of 30 min. Over the smaller fields, in which the cones are concentrated, the sensitivity increases only slightly more than 10 times, and this increase is probably primarily due to the effect of the opening of the iris.

A familiar example is the experience of entering a dark movie theater from brightly sunlit outdoors – you seem *blind*! After groping to a seat, however, you notice that you can begin to detect, dimly, your surroundings. And after a few more minutes you can clearly observe details – you are dark-adapted.

A particular application of the problem of detection of discrete light sources such as stars in competition with the background of the twilight sky is illustrated in Figure 1-4, in which it is shown that fainter and fainter stars can be detected as the background light becomes dimmer. Under excellent sky conditions, stars can be detected to a level between visual magnitudes 5 and 6 at the end of twilight.

The detection of a point source such as a star and of an extended source covering many degrees, such as the gegenschein (to be discussed in Chapter 3) emphasizes the different aspects of visual perception. As described above, the cones are concentrated in the center of the retina and each cone is connected to the brain by a separate nerve; hence vision is very distinct when the light is bright so that the iris closes down and the center of the retina is used. The rods' chemical change during dark adaptation, and their concentration toward the outer part of the retina, permit improved detection of

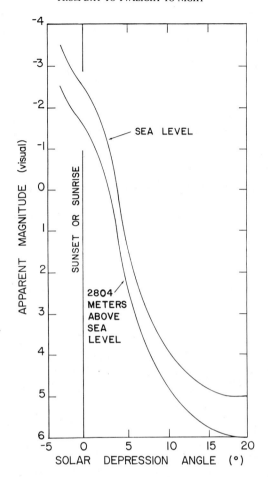

Fig. 1-4. Detectability of stars and planets during twilight. Between sunset and a solar depression angle of 7° the threshold at sea level changes from $m = -2.4$ to $+2.6$, a change of 10^2 in brightness.

faint extended sources by (1) a dark-adaptation period of up to 30 min and (2) preferential use of the outer parts of the retina (averted vision). The detection of stars during twilight is a problem of contrast plus dark adaptation.

1.4. The Night Sky

Everything considered, Earth dwellers are extraordinarily fortunate in that the several separate and distinct components of the light of the night sky can be disentangled, at least to a reasonable extent. We shall explore later some unhappy circumstances that might have made such an analysis of resolving the components of the night-sky light difficult or even impossible.

The several components of the light of the night sky are listed in Table 1-I, together with their physical origins and their coordinate systems. Note that three different coordinate systems are applicable to the several components. This situation is analogous to a case of physical measurement in which three variables are involved, each of which is a function of one of the three coordinates in a Cartesian system. In our case, we have complications in that two of the components, the airglow and the aurora, have strong time dependence. The viewer can effectively eliminate the aurora, however, by avoiding the vicinity of the auroral zone.

A first, gross separation of the components by virtue of their dependence on different coordinate systems is illustrated in Figure 1-5. Several thousands of zenith observations at Haleakala (Island of Maui, Hawaii) have been assembled. The temporal airglow changes have been averaged; the zodiacal light has measurable variations as the ecliptic latitude changes during the year; and the integrated starlight shows two peaks of brightness as the two arms of the Milky Way cross the Haleakala zenith at sidereal times 90° (6 hr) and 295° (19 hr 40 min). In this plot we do not distinguish the minor constituents – the diffuse galactic light and the integrated cosmic light – since their detection calls for a very detailed attack on the data. In Figure 1-6 we show a graphical summary of the distribution of the components of the light of the night sky at a location well away from the auroral zone.

TABLE 1-I

Components of the light of the night sky

Component	Function of		Physical origin
	Position in sky	Other	
Integrated starlight	b, l;[a] galactic coordinates		Stars of our Galaxy
Zodiacal light	$\beta, \lambda - \lambda_\odot$; ecliptic coordinates		Sunlight scattered by interplanetary dust
Airglow	Z, A; horizon coordinates	Local time Latitude Season Solar activity Wavelength Height	Ambient excitation (usually photochemical) of upper-atmosphere atoms and molecules
Aurora		Magnetic latitude Season Solar activity Magnetic activity Wavelength	Excitation of upper-atmosphere atoms and molecules by energetic particles
Diffuse galactic light	b, l		Scattering of starlight by interstellar dust
Integrated cosmic light	b, l	Cosmological red shift Cosmological model	

[a] b = galactic latitude; l = galactic longitude; β = ecliptic latitude; λ = ecliptic longitude; $\lambda_\odot$ = ecliptic longitude of Sun; Z = zenith angle; A = azimuth.

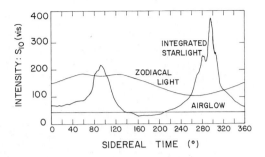

Fig. 1-5. Resolution as a function of sideral time of total zenith intensities at Haleakala (Hawaii) into three components – airglow, zodiacal light, and integrated starlight. In the latitude of Haleakala (about 20°), the ecliptic is in the zenith twice during the year – at 60° (4 hr) and 120° (8 hr) – and thus the zodiacal light shows two slight maxima.

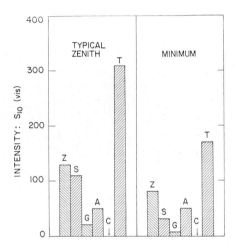

Fig. 1-6. Components of the light of the night sky. Minimum values for zodiacal light, starlight, and the airglow occur at the ecliptic pole, the galactic pole, and the zenith, respectively. Z = zodiacal light; S = integrated starlight; G = diffuse galactic light; A = airglow; C = cosmic light; T = total.

The plan of the book is to discuss the five components of the light of the night sky. We shall concentrate on the visual part of the spectrum, giving brief mention, however, to recent satellite observations in the near-infrared and ultraviolet. In addition to describing the observational facts, we shall discuss some of their implications and interpretations.

References

Ashburn, E. V.: 1952, *J. Geophys. Res.* **57**, 85.
Rayleigh III, John William Strutt, Baron: 1870 (Ms.), 16.
Rayleigh III, John William Strutt, Baron: 1871, *Phil. Mag.*, February.
Rayleigh III, John William Strutt, Baron: 1877, *Theory of Sound* (Macmillan, London).
Rayleigh IV, Robert John Strutt, Baron: 1968 (augmented edition), *Life of John William Strutt, Third Baron Rayleigh, O.M., F.R.S.* (University of Wisconsin Press, Madison).

Lord Rayleigh III
(Courtesy of the Royal Society)

PLATE I   JOHN WILLIAM STRUTT, LORD RAYLEIGH III
                              1842–1919

Lord Rayleigh III is known as 'the scattering Rayleigh', to distinguish him from his son, Lord Rayleigh IV, who is dubbed 'the airglow Rayleigh'. Rayleigh III, however, distinguished himself across a broad spectrum of the science of a century ago. He might also be called 'the sound Rayleigh', as his *Theory of Sound*, published in 1877, is still one of the most respected books on the subject.

Born on November 12, 1842, John William Strutt was descended from a distinguished and well-to-do British family. His grandfather, Colonel John Holden Strutt, had been granted the peerage at the coronation of King George IV as a reward for outstanding public service. He had requested that the title be awarded in the person of his wife, Lady Charlotte Strutt. The name Rayleigh was apparently somewhat arbitrarily chosen, although Colonel Strutt did own some property near the small town of Rayleigh in Essex. His son John James became Lord Rayleigh II on the death of Lady Charlotte in 1836. No scientific interest or talent seems to have been apparent in the family to this time.

The childhood of Rayleigh III was characterized by poor health and lively curiosity. Not until his undergraduate years at Cambridge, however, did his genius for mathematics and science appear. A gifted theoretician, he verified his mathematical results with painstaking experimental work, and one of his greatest laments, both at Cambridge and later in his own home, was for the lack of laboratory equipment. He finally hired a man to constuct items to his specifications for his home laboratory.

More than 100 yr ago, in 1871, Rayleigh III published his classical paper on scattering, 'On the Light from the Sky – Its Polarisation and Colour' (1871), in which he explained the blueness of the day sky. The law governing the effect, stated below, is reproduced from his original handwritten draft (1870, Ms.)

*When light is coscattered by par-
ticles which are very small
compared to ~~the~~ any of the wave lengths
the ratio of the amplitudes
of the vibrations of the scattered
and incident light varies in-
versely as the square of the
wave lengths, & the intensity
of the lights themselves as the
inverse fourth power.*

From 1879 through 1884, Rayleigh held a Cambridge professorship, but then returned to his home in Terling to conduct his experiments and develop his theories. During this period he published a paper in *Nature* on the subject of mechanical flight, and in 1900 he lectured on 'The Mechanical Principles of Flight'.

In 1887, Rayleigh accepted the Professorship of Natural Philosophy at the Royal Institution, where he presented lectures illustrated with experiments of his own devising – on such varied subjects as sound and vibration, optics, electricity and magnetism, light and color, the mechanical properties of bodies, matter at rest and in motion, the forces of cohesion, and polarized light. He resigned the Professorship in 1905 but gave occasional lectures after that time.

Throughout his life he was not only friend and co-worker with the major scientists of his period – A. J. Ångstrom, Robert Boyle, Niels Bohr, Charles Darwin, Albert Einstein, Michael Faraday, Willard Gibbs, Hermann Helmholtz, Sir John Herschel, Lord Kelvin, Lister, Clerk Maxwell, A. A. Michelson, Sir William Ramsey, Ernest Rutherford, Sir Frederick Stokes, John Tyndall, and Sir J. J. Thomson – but also with many high-level social, intellectual, and political figures, such as Winston S. Churchill, Joseph Chamberlain, Arthur Balfour, and Lord Salisbury.

His later years were characterized by a broadening of his activities into more public work – local political administration, matters of public concern – but he remained devoted to science to the end of his life, in 1919.

Rayleigh's old friend Sir J. J. Thomson presented the memorial address in Westminster Abbey; a few excerpts from his address may summarize Rayleigh III most effectively:

"Among the 446 papers... there is not one that is trivial... and among that great number there are scarcely any which time has shown to require correction.... The first impression we gain... is the catholicity of Lord Rayleigh's work.... Lord Rayleigh took physics for his province and extended the boundary of every department of physics... in the majority of his writings we have a combination of mathematical analysis and experimental work, and his papers... afford the best illustration of the true co-ordination of these two great branches of attack on the problems of nature. The physical ideas direct the mathematical analysis into the shortest and most appropriate channels, while the mathematics gives precision and point to the physics." (Rayleigh IV, 1968, pp. 309–310.)

Having made unique scientific contributions through about 55 of his 77 years, Lord Rayleigh III merited the Memorial in Westminster Abbey, on which the inscription reads:

"JOHN WILLIAM STRUTT, O.M., P.C.

3rd Baron Rayleigh

Chancellor of the University of Cambridge, 1908–1919

President of the Royal Society, 1905–1908

An Unerring Leader in the Advancement of Natural Knowledge."

(Rayleigh IV, 1968: 374)

CHAPTER 2

STAR COUNTS AND STARLIGHT

In Chapter 1 we found that the unaided eye could distinguish stars to the 6th visual magnitude from a mountain top and to the 5th magnitude from sea level. Over the entire sky there are 4850 stars brighter than magnitude 6 and 1620 brighter than magnitude 5. These are manageable numbers; but with telescopic aid the number of stars either visible or photographable increases enormously. For example, there are

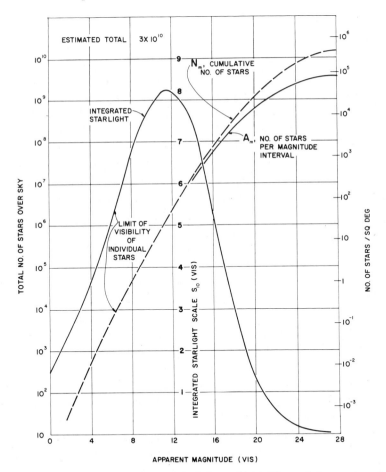

Fig. 2-1. Plots of: (a) number of stars per magnitude interval (A_m), (b) cumulative number of stars (N_m), and (c) integrated starlight (J_m) per magnitude interval as a function of apparent visual magnitude. (Based on MW301, averaged over the whole sky.)

10^6 stars brighter than visual magnitude 11.2 and 10^9 brighter than magnitude 20. A definitive count is possible for the brighter stars, but for the fainter ones counting has been completed only for samples.

The entire sky has been photographed using telescopes designed to include large regions in good focus. The National Geographic Society-Palomar Observatory Sky Survey consists of 1036 separate reproductions from original plates made with the Palomar 48-in. Schmidt telescope for that part of the sky photographable from the Palomar latitude. Stars as faint as photographic magnitude 20 (visual magnitude 18.8) appear on the reproductions. Figure 2-1 indicates that there are of the order of 5×10^8 stars brighter than visual magnitude 18.8. Note, however, that this is less than 2% of the total number of stars in our Galaxy.

The task of counting and tabulating this vast reservoir of information is formidable. Historically, it has been approached in two ways – (1) by starting with the less-numerous bright stars, and (2) by concentrating on a few selected, and hopefully representative, regions of the sky (Figure 2-2).

In the 18th century, William Herschel – a German-born musician living in England – decided to abandon his first love, music, and take up a new enthusiasm, the building and use of telescopes. He was a dedicated and careful observer and, with the help of his sister Carolina, he did considerable star 'gaging' (which we now call star counting), compiling data on an essentially statistical basis and developing an imaginative approach to an understanding of galactic structure. From the results of his early work grew the first concept of the morphology of the star system known as our Galaxy (Figure 2-3a). (See Appendix 2-A for an example of star gaging by Herschel and a quantitative comparison with modern star counting.) Plate II presents Herschel, the outstanding innovator, with a brief story of his life and scientific contributions.

2.1. Plan of Selected Areas

An extension and expansion of Herschel's approach occurred about a century after Herschel's period of activity, in 1906, when J. C. Kapteyn initiated an international program of systematic star counting based on his Plan of Selected Areas. Kapteyn chose 206 regions in the sky for a concentrated study of brightnesses, proper motions, trigonometric parallaxes, positions, spectral classifications, absolute magnitudes, and radial velocities of the stars (including variables) in each area.

The distribution of these areas over the sky is systematic in the equatorial coordinate system. Areas 1 and 206 are near the north and south celestial poles respectively. Areas 2 through 205 are distributed as follows: 12 areas at $\pm 75°$ declination (six at plus and six at minus declination), 24 areas at $\pm 60°$, 48 at $\pm 45°$, 48 at $\pm 30°$, 48 at $\pm 15°$, and 24 at $0°$ (Figure 2-4). Figure 2-3b shows a galactic model based on early work on Kapteyn's Selected Areas. An account of Kapteyn and his work is presented in Plate III.

Through the 20th century, since he initiated the Selected Areas concept in 1906, productive activity on Kapteyn's method has continued steadily.

Galactic Longitude = 49°, Galactic Latitude = +74°

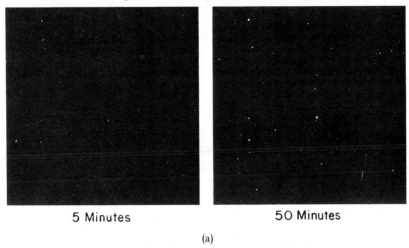

5 Minutes   50 Minutes

(a)

Galactic Longitude =27°, Galactic Latitude =-4°

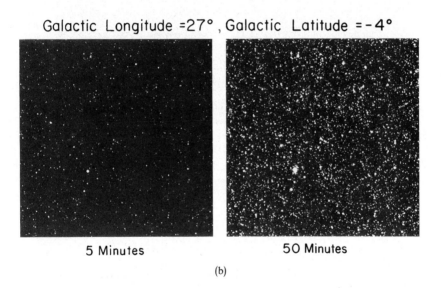

5 Minutes   50 Minutes

(b)

Fig. 2-2. Photographs of star-poor (2-2a) and star-rich (2-2b) fields, at short (5 min) and long (50 min) exposure times. The 2-2a field is about 16° from the North galactic pole ($\alpha = 14$ h 01 m; $\delta = +30°5$) and is Selected Area 58, labelled A on Figure 2-4. The star-rich field is 4° from the galactic equator, in the constellation Scorpius ($\alpha = 18$ h 55 m; $\delta = -7°0$). The field size of each photograph is 16′ by 16′, or 0.07111 sq deg. The effect on the number of stars visible as the exposure time is increased is evident by comparing the photographs left and right. Counts of the number of stars on the working prints gave 16 for a(5), 44 for a(50), 930 for b(5), and 4500 for b(50). Referring these counts to the quantity N_m (number of stars brighter than magnitude m per sq deg – see p. 17), we get $\log N = 2.4$ for a(5), 2.8 for a(50), 4.1 for b(5), and 4.8 for b(50). The photographic limiting magnitude is about 16 for the 5 min exposures and about 18 for the 50 min exposures. The photographs were made by B. J. Bok with the $f/9$ Cassegrain arrangement of the Steward Observatory 90 in. reflector at Kitt Peak.

(a)

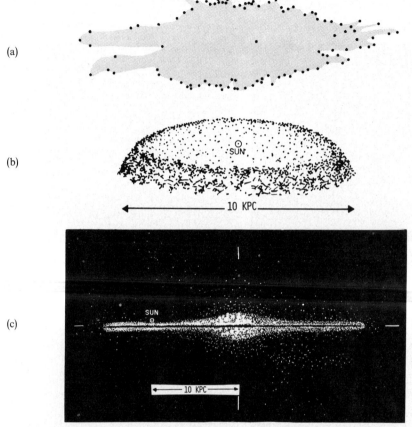

(b)

(c)

Fig. 2-3.   Developing concepts of our Galaxy according to: (a) Herschel, (b) Kapteyn, and (c) present (seen edge-on).

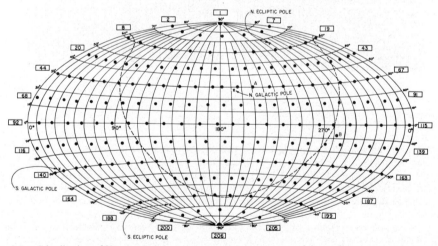

Fig. 2-4.   Distribution of Kapteyn's Selected Areas, shown on equatorial coordinates; the galactic equator is shown as the dashed line. Region A (near the galactic pole) is the star-poor field of Figure 2-2a; region B (near the galactic equator) is the star-rich field of Figure 2-2b.

2.2. Modified Star Counting

An understanding of the morphology of our Galaxy has been significantly advanced during the last two decades by a continuation of the star-counting technique as a probe. The counting approach has departed from the gross statistical procedures characteristic of the earlier work and has concentrated on detailed studies in which a depth inference in the line of sight is made from the observed distribution of the brightnesses of stars projected against the celestial sphere, together with measurements of their spectral characteristics *.

The analysis is complicated by the fact that the apparent brightness of a star of

Fig. 2-5. Spiral Galaxy Messier 51. Our Galaxy would probably look similar to M51 if seen face-on from a comparable distance. The approximate location of our Sun would be about two-thirds of the way out from the center, in one of the spiral arms (see Figure 2-6).

* Herschel was handicapped in his early attempts to obtain even an approximation to a depth analysis because, in the absence of any knowledge as to the very wide distribution of intrinsic stellar brightnesses as revealed by spectroscopic means, he used the gross oversimplification that 'brightness equals nearness'.

given intrinsic luminosity depends on both its distance and its partial obscuration by
dust between star and observer. (An assist has come from the radio astronomers,
who have taken advantage of the fact that radio waves penetrate through the ubiquitous dust in our Galaxy much more efficiently than do the optical radiations.)

An example of such an analysis is the present picture of our Galaxy as a spiral
similar to the many external spirals that have been photographed by our major telescopes (Figure 2-5). A somewhat schematic illustration of three spiral arms that have
been identified in the part of our Galaxy near our Sun is shown in Figure 2-6. The Sun

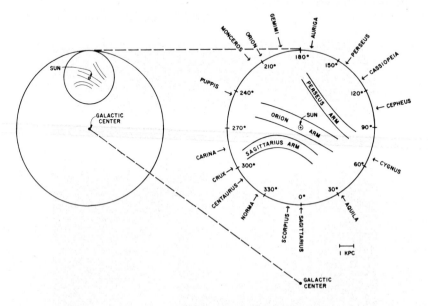

Fig. 2-6. Spiral structure of the portion of the Galaxy near the Sun, based on optical and statistical studies.
The page is in the galactic plane. Galactic longitudes, l^{II}, are indicated around the circumference of a circle
centered on the Sun, together with some of the constellations visible to a solar-terrestrial observer. The
direction toward the galactic center (10 kpc distant) is at $l^{II}=0°$ in the constellation of Sagittarius.

is embedded in one of the spiral arms – the so-called Orion arm. The Sagittarius arm
and the Perseus arm are as near as 1 and 2 kpc (kiloparsecs), respectively, and have
been delineated out to several kpc. For reference, the distance of our Sun from the
center around which the Galaxy revolves is about 10 kpc.

The use of such a probing of an arm in our spiral-type Galaxy has led to major
contributions to our knowledge in recent years by Bart J. Bok, who is still exploring
this productive technique (Plate IV).

2.3. Distribution of Integrated Starlight over the Sky

In Table 2-I and Figure 2-1, we show some statistical results based on a study of
Selected Areas 1 to 139 (Seares *et al.*, 1925). The data are in visual magnitudes and
refer to the mean over the sky. The Selected Areas chosen included the zone as far

TABLE 2-I

Summary of star counts by visual-magnitude intervals; mean over the sky (from MW301[a])

m(vis)	$\log A_m$	$\Delta \log A_m$	J_m $(S_{10}$(vis))	J_m(cum) $(S_{10}$(vis))	$\log N_m$	$\log T_m$
0	−3.821		1.51	105.20		
1	−3.301	0.520	1.99	103.69		
2	−2.807	0.494	2.47	101.70		
3	−2.326	0.481	2.98	99.23		
4	−1.841	0.485	3.62	96.25	−1.888	2.727
5	−1.362	0.479	4.34	92.63	−1.405	3.210
6	−0.897	0.465	5.05	88.29	−0.929	3.686
7	−0.431	0.466	5.88	83.24	−0.461	4.154
8	+0.023	0.454	6.65	77.36	+0.001	4.616
9	0.466	0.443	7.34	70.71	0.453	5.068
10	0.897	0.431	7.90	63.37	0.895	5.510
11	1.312	0.415	8.17	55.47	1.324	5.939
12	1.714	0.402	8.20	47.30	1.740	6.355
13	2.097	0.383	7.89	39.10	2.141	6.756
14	2.461	0.364	7.26	31.21	2.525	7.140
15	2.805	0.344	6.38	23.95	2.891	7.506
16	3.120	0.315	5.25	17.57	3.235	7.850
17	3.411	0.291	4.08	12.32	3.557	8.172
18	3.677	0.266	3.00	8.24	3.856	8.471
19	3.916	0.239	2.07	5.24	4.132	8.747
20	4.131	0.215	1.35	3.17	4.385	9.000
21	4.318	0.187	0.83	1.82		
22	4.481	0.163	0.48	0.99		
23	4.618	0.137	0.26	0.51		
24	4.729	0.111	0.14	0.25		
25	4.815	0.086	0.06	0.11		
26	4.875	0.060	0.03	0.05		
27	4.910	0.035	0.01	0.02		
28	4.919	0.009	0.01	0.01		(10.48)

[a] Entries corresponding to m(vis) > 20 are based on a quadratic extrapolation of the data in MW301 (Seares et al., 1925).

south as declination −15°, which can be observed at Mt. Wilson during meridian passage at a zenith angle of 49°. (The next zone to the south, at a declination of −30°, has a zenith angle of 64° for meridian passage; although observable, this angle is a little low in the sky for accurate photometry.) Much of the bright Milky Way in the southern sky is not included in the Mt. Wilson study, but the over-all results can be profitably examined.

In Figure 2-1 we have used three scales for the ordinate. To the left is indicated the total number of stars over the sky; to the right, the number per square degree. Both scales can be referred to either the cumulative number of stars, N_m (number of stars brighter than magnitude m), or to the number per magnitude interval, A_m (number of stars between magnitudes $m-\frac{1}{2}$ and $m+\frac{1}{2}$). The central scale applies to the integrated starlight, J_m, the brightness in S_{10}(vis) units from the stars between magnitudes $m-\frac{1}{2}$ and $m+\frac{1}{2}$. The relationship between A_m and J_m is

$$J_m = A_m \cdot 10^{-0.4(m-10)}.$$

The plot of J_m is of direct interest to the topic of this book, since the summation under the curve gives the mean contribution of starlight to the light of the night sky. J_m goes through a maximum at m(vis)$=12$. The position of the maximum depends on the point of inflection in the $\log A_m$ vs m curve, where $(\varDelta \log A_m)/\varDelta m = 0.4$. When $(\varDelta \log A_m)/\varDelta m$ is greater than 0.4, the number density of stars increases more rapidly than their brightness decreases, and J_m increases with m; when $(\varDelta \log A_m)/\varDelta m$ is less than 0.4, J_m decreases with m. The maximum occurs for $(\varDelta \log A_m)/\varDelta m = 0.4$.

The summation of J_m (column 5 in Table 2-I) is made from the faint end, since we are interested in evaluating the background starlight with respect to observers for whom stars of different brightnesses (magnitudes) are discretely observed and hence do not contribute to the 'interstellar' background. For example, the observer who scans the sky without a telescope will see discrete stars to the 5th or 6th mag., and the background starlight *on the average* is about 90 S_{10}(vis) units. An observer using a 4-in. telescope sees stars as faint as magnitude 12, and the background of unresolved stars will be only some 47 S_{10}(vis). With a 100-in. telescope, stars of magnitude 19 are discernible, and the background starlight is reduced to about 5 S_{10}(vis).

2.4. Distribution of Starlight over the Sky

In the preceding paragraphs we have considered the average background of light from unresolved stars. The departure from the average is of interest to the investigator of galactic structure but also to us, because we are concerned with the distribution of starlight over the sky.

As mentioned earlier, the fact that the galactic system concentrates the stars toward the galactic equator (Milky Way) facilitates the disentanglement of integrated starlight from the other major constituents of the night-sky light. We plan to pursue this matter with reference to two compilations of star counts – (1) the Mt. Wilson publication 301 (Seares *et al.*, 1925) previously mentioned, and (2) Groningen Astronomical Observatory publication 43 (van Rhijn, 1925). The former is based on Selected Areas 1 through 139, the latter on the totality of the Areas.

In Figure 2-7 we show plots of J_m for galactic latitudes $b=0°$ and $80°$, from the two published sources. The data are assembled in Table 2-II. Figure 2-8 displays the relationship between J_m(cum) and galactic latitude for the mean of GR-43 and MW-301 (from columns 4 and 8 of Table 2-III). In general, the integrated starlight is about nine times as bright in the Milky Way as near the galactic pole. The fact that the Milky Way does not appear to the terrestrial-bound eye with such a contrast illustrates that there is a dilution of the contrast by other sources (zodiacal light and nightglow) in the night-sky light.

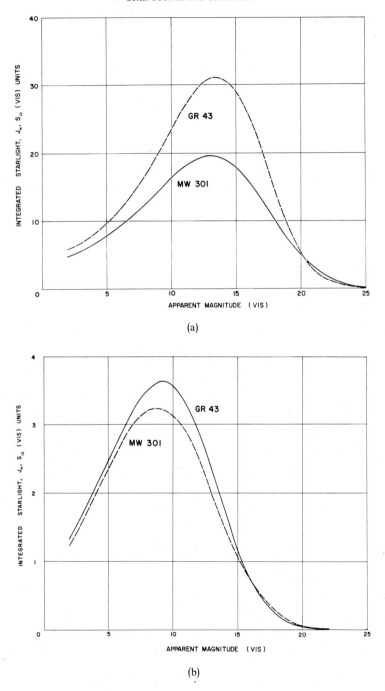

Fig. 2-7.   Plots of integrated starlight vs visual magnitude for the region of the galactic equator ($b=0°$) and a region near the galactic pole ($b=80°$). The results are shown for the two independent investigations, MW301 and GR43. Note the ten-fold differences in the scales of the ordinates between the two plots. Plot (a) is for $b=0°$ and plot (b) for $b=80°$.

TABLE 2-II

Comparison of star-count data from GR43 and MW301[a].

m(vis)	b=0°				b=80°			
	J_m		J_m(cum)		J_m		J_m(cum)	
	GR43	MW301	GR43	MW301	GR43	MW301	GR43	MW301
2	5.91	4.61	341.63	238.12	1.33	1.23	37.29	34.04
3	6.73	5.42	335.72	233.51	1.67	1.59	35.96	32.81
4	7.88	6.42	328.99	228.09	2.07	1.98	34.29	31.22
5	9.42	7.65	321.11	221.67	2.49	2.38	32.22	29.24
6	11.37	9.10	311.69	214.02	2.90	2.75	29.73	26.86
7	13.81	10.76	300.32	204.92	3.26	3.04	26.83	24.11
8	16.70	12.60	286.51	194.16	3.52	3.22	23.57	21.07
9	19.98	14.52	269.81	181.56	3.63	3.25	20.05	17.85
10	23.45	16.37	249.83	167.04	3.56	3.12	16.42	14.60
11	26.77	17.98	226.38	150.67	3.29	2.84	12.86	11.48
12	29.47	19.12	199.61	132.69	2.85	2.44	9.57	8.64
13	31.03	19.58	170.14	113.57	2.30	1.98	6.72	6.20
14	30.96	19.21	139.11	93.99	1.73	1.51	4.42	4.22
15	29.00	17.93	108.15	74.78	1.19	1.08	2.69	2.71
16	25.24	15.84	78.15	56.85	0.74	0.72	1.50	1.63
17	20.20	13.15	53.91	41.01	0.42	0.44	0.76	0.91
18	14.68	10.19	33.71	27.86	0.21	0.25	0.34	0.47
19	9.55	7.32	19.03	17.67	0.09	0.13	0.13	0.22
20	5.46	4.82	9.48	10.35	0.03	0.06	0.04	0.09
21	2.66	2.88	4.02	5.53	0.01	0.02	0.01	0.03
22	1.05	1.54	1.36	2.65		0.01		0.01
23	0.29	0.72	0.31	1.11				
24	0.02	0.29	0.02	0.39				
25		0.09		0.10				
26		0.01		0.01				

[a] Entries corresponding to m(vis) > 20 for MW301 (Seares et al., 1925) and m(vis) > 17 for GR43 (van Rhijn, 1925) are based on a quadratic extrapolation of the published data.

2.5. Photometric Map of the Sky Based on Star Counts

In spite of the dedicated efforts of many astronomers in counting the stars in Selected Areas, the fraction of the sky so scrutinized is very small indeed. In the Mt. Wilson program of 139 Selected Areas, 65 683 stars were measured (an average of 472 per Area) to approximately photographic limiting magnitude 18 (16.94 visual magnitude). The total number of stars over the sky to this magnitude is about 3×10^8; thus the sample measured was 0.022% of the total.

Another approach to the size of the sample in the program is by way of the total area covered. The angular area measured on each plate was 0.1154 sq deg. For 139 Areas, the percent area measured was thus

$$(139 \times 0.1154 \times 100)/41\,253 = 0.039\%.$$

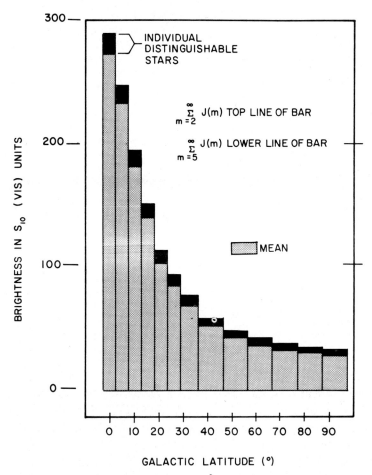

Fig. 2-8.   Integrated starlight vs galactic latitude (mean of MW301 and GR43). The tops of the rectangles represent the total integrated starlight to magnitude 2 – the solid black portion; the shaded region corresponds to the total integrated starlight from stars fainter than those discernible to the eye.

From such a small sample we can hope to get only very crude morphological details of our galactic system and equally crude information on the photometric structure of the Milky Way. The uncertainty due to the smallness of the sample is compounded by the significant discrepancy between the two starcount reports quoted in this chapter.

The GR-43 listing is in the form of tables of the quantity $\log N_m$ from m(pg) 6 to 18 for every $10°$ of galactic longitude l and for galactic latitudes b of $0°$, $\pm 2°$, $\pm 5°$, $\pm 10°$, $\pm 15°$, $\pm 20°$, $\pm 30°$, $\pm 40°$, $\pm 50°$, $\pm 60°$, $\pm 70°$, and $\pm 80°$ (10 296 entries for 792 regions in the sky). Since the study was based on 206 Selected Areas, it is obvious that considerable interpolation entered into the tabulated results.

In order to delineate the details in the regions of low galactic latitude, it is desirable to use photometric measurements rather than deduced brightnesses based on

TABLE 2-III

Relationship between J_m(cum) and galactic latitude for two different m_0,[a]
based on GR43 and MW301 data

b	$J_m(\text{cum}) = \sum\limits_{m=m_0}^{\infty} J_m = J$							
	$m_0 = 5$				$m_0 = 2$			
	GR43	MW301	Mean	Relative	GR43	MW301	Mean	Relative
0	321	222	271.5	9.20	342	238	290	8.53
5	263	202	232.5	7.88	279	217	248	7.29
10	195	167	181	6.14	209	181	195	5.73
15	150	130	140	4.75	161	142	151.5	4.46
20	107	100	103.5	3.51	117	110	113.5	3.34
25		81	(82.5)	2.90		90	(92)	2.71
30	71	68	69.5	2.36	78	76	77	2.26
40	54	51	52.5	1.78	60	58	59	1.74
50	43	42	42.5	1.44	49	48	48.5	1.43
60	38	36	37	1.25	43	42	42.5	1.25
70		32	(33)	1.12		37	(38)	1.12
80	32	29	30.5	1.03	37	34	35.5	1.04
90		28	(29.5)	1.00		33	(34)	1.00

[a] m_0 = cutoff magnitude.

smoothed star-count data. In Table 2-IV we have assembled the brightness of the entire 'astronomical' sky based on two published sources. For galactic latitude $< 30°$ we have taken measured brightnesses from Smith *et al.* (1970); for galactic latitudes $\geqslant 30°$ we used the conversion of the star-count data of Groningen 43 according to Roach and Megill (1961). Plots of the brightness of the Milky Way in two planes (Figures 2-9 and 2-10) illustrate its photometric structure, which is a result of our position in the Galaxy (compare with Figure 2-6).

2.6. The Total Light from the Stars

If the stars were uniformly distributed over the sky and diffused so that they were not individually discerned, they would produce a general glow that, according to Table 2-I and Figure 2-8 would be equivalent to 105 10th-mag. (visual) stars for each square degree of sky. In the hemisphere above the horizon there are 20 626 sq deg, so that the total illumination from the sky is equivalent to 2.17×10^6 stars of visual magnitude 10.

Since a 10th-mag. star is about 100 times fainter than the faintest star visible to the average observer, alternative ways of expressing this are of interest. If, over the hemisphere, there were 51 stars as bright as the nearby and intrinsically bright star Sirius ($m = -1.58$), the total illumination would be the same as is now produced by the 30×10^9 stars in our Galaxy (Table 2-V). This is a dramatic indication of the facts that

Integrated starlight

l \ b	0	−2	−5	−10	−15	−20	−30	−40	−50	−60	−70	−80	80	70	60	50	40	30	19	15	10	5	2	0
0	380	280	330	260	130	100	65	51	43	38	34	31	33	34	39	49	64	91	240	230	500	800	540	380
10	470	330	340	240	170	100	67	52	42	37	33	31	34	34	39	49	64	94	270	380	620	560	540	470
20	330	210	150	160	140	100	72	54	42	37	33	31	34	35	39	47	61	91	220	340	420	350	330	330
30	270	150	160	240	170	100	79	56	42	36	32	30	34	36	39	46	59	84	200	150	160	330	410	270
40	200	210	200	190	130	100	85	58	43	36	32	29	35	36	39	46	57	76	170	220	240	310	290	200
50	210	210	240	170	120	100	89	60	43	36	31	29	34	37	39	45	54	68	140	160	280	370	290	210
60	290	310	300	220	130	90	88	59	43	35	31	28	35	37	39	43	51	64	140	300	290	380	310	290
70	350	430	370	230	130	90	84	57	43	35	30	28	34	37	38	41	48	62	120	140	270	290	290	350
80	250	260	270	210	150	90	76	55	42	35	30	28	34	36	36	39	49	65	110	180	280	300	280	250
90	340	320	250	190	160	90	68	53	41	34	30	28	33	35	35	37	49	70	140	130	240	300	260	340
100	280	220	180	180	120	90	61	50	41	33	29	28	33	33	34	37	50	76	140	180	210	330	330	280
110	270	240	190	170	120	90	57	47	40	34	30	28	32	32	33	37	51	79	140	150	190	250	240	270
120	280	220	170	140	120	90	54	45	40	34	31	28	32	31	32	38	51	75	130	130	180	290	320	280
130	250	200	160	140	110	100	53	45	39	34	31	29	32	31	32	38	50	68	140	170	180	260	290	250
140	180	170	160	120	120	100	54	44	39	34	31	30	32	31	33	38	48	58	190	160	180	210	180	180
150	160	150	130	120	120	110	54	43	38	34	31	30	32	31	34	38	44	50	160	210	190	180	190	160
160	180	170	140	150	130	110	55	44	38	33	31	30	32	32	34	37	41	46	190	180	180	160	180	180
170	230	220	170	210	130	120	57	46	36	33	31	29	32	32	34	37	40	43	180	240	150	190	200	230
180	220	230	220	180	160	140	58	43	35	32	29	28	32	32	34	36	41	45	170	180	160	210	220	220
190	220	250	230	210	170	160	59	43	33	30	29	28	32	32	33	36	43	51	140	170	150	220	230	220
200	220	220	280	170	160	140	61	43	33	30	27	27	32	32	34	38	46	59	100	160	140	260	220	220
210	220	220	200	180	150	110	63	42	33	29	27	27	32	31	34	40	51	69	100	140	140	170	200	220
220	240	250	220	160	130	100	64	43	33	29	27	27	32	32	35	42	54	77	110	130	200	200	220	240
230	250	250	220	150	100	90	64	43	34	29	29	28	33	33	35	43	57	81	110	140	230	240	250	250
240	330	310	240	150	100	90	62	44	35	30	30	28	33	33	36	43	57	81	110	140	210	370	290	330
250	270	280	230	160	120	90	63	46	37	31	31	30	34	34	37	42	55	78	110	140	200	260	270	270
260	230	220	200	150	110	90	63	49	39	33	32	30	34	35	37	41	53	72	110	150	220	280	270	230
270	230	210	210	170	120	90	65	51	41	34	32	30	34	36	37	41	50	68	120	190	260	260	230	230
280	370	350	260	190	120	100	71	53	43	35	32	30	35	36	38	41	49	65	130	200	280	440	400	370
290	640	560	340	210	140	100	76	56	45	36	32	29	35	36	39	43	50	63	120	170	270	360	480	640
300	660	570	530	220	160	110	81	59	46	37	32	29	35	40	40	45	52	64	70	150	220	380	560	660
310	570	530	370	210	190	130	84	58	46	37	31	29	34	36	42	48	56	65	140	160	250	330	460	570
320	260	280	270	230	210	140	82	58	46	37	32	29	34	36	42	46	59	68	140	250	320	420	360	260
330	390	280	270	260	190	120	76	56	45	37	33	30	33	35	42	51	62	73	150	210	310	460	610	390
340	300	280	140	170	160	110	70	54	44	38	33	30	34	35	42	53	64	80	180	240	340	330	300	300
350	360	350	360	290	140	110	66	51	43	38	34	30	33	35	41	52	65	87	240	310	400	560	700	360
0	380	280	330	260	130	100	65	51	43	38	34	31	33	34	39	49	64	91	240	230	500	800	540	380

Brightness J in S_{10}(vis) units (lower b header, left→right): 0, 2, 5, 10, 15, 19, 30, 40, 50, 60, 70, 80, −80, −70, −60, −50, −40, −30, −20, −15, −10, −5, −2, 0

[a] $|b| \geq 30°$ from Roach and Megill (1961). $|b| < 30°$ from Smith et al. (1970). J at galactic pole = 28.

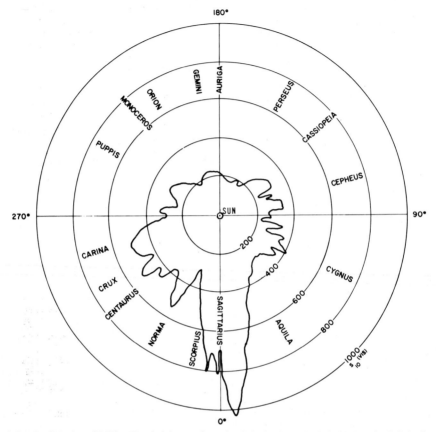

Fig. 2-9. Radial plot of Milky Way brightness for galactic latitude $-3°$. Galactic longitude is indicated
increasing from $0°$ (toward the galactic center) in a counterclockwise sense.

most of the stars in our Galaxy are (1) much more distant than Sirius (2.7 pc $=8.8$ l-yr)
and (2) intrinsically fainter.

The average surface brightness of the starlit sky can be expressed in terms of the
surface brightness of the Sun, with its surface area of 0.2233 sq deg. In S_{10}(vis) units,
the Sun's mean surface brightness is 2.18×10^{15}. Thus the surface brightness of the
starlit sky is $105/2.18 \times 10^{15} = 0.5 \times 10^{-13}$ times that of the Sun.

As the realization grew that the stars are suns, a speculation arose in thoughtful
minds concerning the possibility that the sky might, under some assumptions, approach a much higher brightness. Following an earlier suggestion by the astronomer
Halley, Wilhelm Olbers, an Austrian physician, made the following statement in 1826:

If there really be suns in the whole of space, and to infinity, and if they are placed at equal distances from
each other, or grouped into systems like that of the Milky Way, their number must be infinite, and the whole
vault of heaven should appear as bright as the Sun; for every line which may be supposed to emanate from
our eye towards the sky, would necessarily meet a fixed star, and thus every point of the sky would bring
to us a ray of sideral [sic.], or which is the same thing, of solar light. (Olbers, 1826.)

The discrepancy (or paradox, as it is sometimes called) between the speculative idea

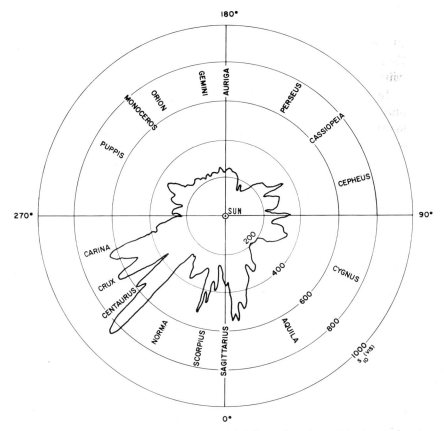

Fig. 2-10.   Radial plot of Milky Way brightness for galactic latitude 0°.

TABLE 2-V

Equivalent expressions for starlit-sky brightness

Reference object	Apparent magnitude	Number to equal one hemisphere of starlight
10th-mag. star	10.0	217×10^4
1st-mag. star	1.0	590
Sirius	− 1.58	51
Venus (at brightest)	− 4.3	4
Full Moon	−12.5	22×10^{-5}
Sun	−26.72	445×10^{-12}

of Olbers and the observed 105 S_{10} (vis) is about 13 powers of 10. The Olbers' premise of an infinite number of stars extending in a uniform distribution to infinity is obviously not supported by the observations. The idea of continuing the discussion with galaxies replacing stars in the statement, however, has excited the interest of some cosmologists, and we shall return to this idea in the last chapter.

2.7. Our Felicitous Location in the Galaxy

We have called attention to our peripheral and somewhat isolated position in the Galaxy. How would the astronomical sky appear if we inhabited a planet associated with a star in (1) the center of our Galaxy, or (2) the center of a dense globular cluster?

In or near the galactic center, the general sky would be some 60 times as bright as our terrestrial sky (Table 2-VI). The probable presence of interstellar dust would have the effect of scattering the general starlight and would interdict any optical inquiries to the outside.

TABLE 2-VI

The night skies of different locations

Location	S^a	$\bar{S}$		S_1		S_{10}(vis)	m(vis)
	(stars pc^{-3})	(pc)	(l-yr)	(pc)	(l-yr)		
Earth-Sun	0.1	2.15	7.03	1.19	3.89	105	–
Galactic center	45	0.28	0.92	0.16	0.51	6 000	−4.1
Center of dense globular cluster	1000	0.1	0.33	0.055	0.18	48 000	−6.5

[a] S is star density
   $\bar{S}$ is mean separation of stars $= \sqrt[3]{S^{-1}}$
   S_1 is the closest star neighbor $= 0.554\,\bar{S}$

S_{10}(vis) is astronomical sky brightness
m(vis) is apparent visual magnitude of nearest star (assumed same luminosity as Sun)

A central position in a dense globular cluster would afford a brilliant sky indeed, but one in which the Milky Way would be photometrically overwhelmed by the many bright close stars over the sky. The nearest star neighbor, only about 66 light days distant, would be a brilliant star. If this neighbor happened to be a star like the Sun, its apparent magnitude would be about −6.5. The bright sky would discourage optical explorations beyond the cluster – perhaps the astronomers would not even realize that they were members of a galactic system. Of course, the local SPACE organizations would be busy raising appropriations for interstellar travel, since there is high likelihood that intercommunication by radio would occur relatively early in the development of a technological civilization and messages could be answered in about 132 days (terrestrial).

We consider that our particular location in the Galaxy is a happy one, even though we are certainly not suffering from stellar overcrowding. The advantage of a moderately luminous sky, which encourages rather than inhibits astronomical exploration, outweighs the loneliness of our position of relative isolation in the outer portion of our Galaxy.

Appendix 2-A. Star Gaging (Herschel) and Star Counts

A comparison of some of the numerical data of William Herschel's visual star gaging with modern star-counting results based on photographs will illustrate the nature of the statistics of stellar distributions over the sky.

Based on the discussion of Herschel's work in Hoskin (1963), we give a summary sample of some of his gaging results with the use of a telescope of 18.7-in. aperture and field of view of 15′ diameter (0.049 sq deg area).

Galactic latitude (deg)	Gage	N (sq deg)$^{-1}$
~ 0	79.0	1610
26	18.6	379
29	13.6	277
32	12.1	247
36	10.6	216
41	10.6	216
43	9.4	192
60	3.6	73
64	3.1	63
72	3.8	77
73	4.0	82

The numbers in the last column are the numbers of stars referred to a field of 1 sq deg and are therefore eligible for comparison with the quantity N_m (number of stars per sq deg brighter than magnitude m).

We show such a comparison in Figure 2-11, in which the solid lines refer to highly smoothed means from Groningen 43 (van Rhijn, 1925). We note that this sample of Herschel's observations suggests that his effective limiting magnitude was about 14 (visual). This value is reasonable considering the nature of his observing procedure,

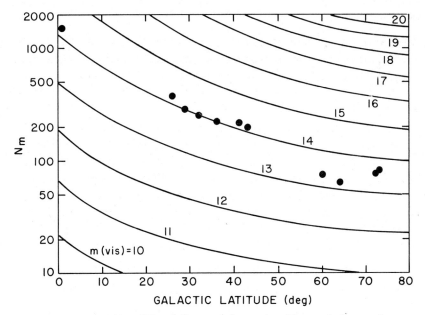

Fig. 2-11.   Comparison of Herschel's star-gaging results with smoothed means from Groningen 43 modern star counts.

which necessarily involved a 'sliding field' instead of a stable field as in a modern automated telescope.

Appendix 2-B. Photometric Units in Surface Photometry

In this chapter we have used almost exclusively as a unit for expressing night-sky 'brightness', S_{10}, the equivalent number of stars of magnitude 10 per sq deg. In Chapter 4 we shall use another unit, the rayleigh, for indicating the quantal output of upper atmosphere emissions. To the general reader these units are probably unfamiliar and we shall discuss them here.

A. *The S_{10} Unit*

The use of the square degree as the unit of solid angle in the S_{10} unit follows historically from the fact that the square degree is the reference in stellar statistics (see Appendix 2-A, for example). There are 41 253 sq deg over a sphere, 3283 in a steradian – 1 sq deg is thus 0.000 305 ster. The choice of magnitude 10 rather than some other – such as 5, which is occasionally used – is arbitrary.

Except for the bright zodiacal light and the brightest part of the Milky Way, brightnesses in S_{10} units are expressible with three digits (less than 999), which is convenient in practical computations.

We have used either the astronomical visual or photographic magnitude scale in this book, avoiding the difficulties that arise in the use of the S_{10} unit in connection with the multicolor observations of modern photoelectric photometry.

B. *The Rayleigh Unit*

In dealing with the upper atmosphere, the source of the radiation is emitting atoms or molecules. If the radiance is expressed in any of the usual photometric units, it is necessary to convert the reading into the number of quanta involved in order to associate the observation with the physics of atmospheric processes. Lord Rayleigh IV recognized this problem in his pioneering observations of O I 5577 Å and reported that the radiation referred to the zenith was '... 1.81×10^{12} atomic transitions per second per square meter' (Rayleigh IV, 1930). His observations were made in London out of a window that did not permit direct observation of the zenith. The absolute calibration was accomplished by pointing the photometer toward a magnesium oxide screen illuminated by a laboratory standard light. In recognition both of Rayleigh's skill in making the absolute measurement of the 5577 Å intensity and of his insight in expressing his results in a form palatable to the upper-atmosphere physicist, a unit called the rayleigh has been defined –

If the surface brightness, B, is measured in units of 10^6 quanta cm^{-2} s^{-1} ster^{-1}, then in rayleighs, R, the surface brightness is $4\pi B$ (Hunten *et al.*, 1956).

Thus Rayleigh's measurement comes out as 181 R, in substantial agreement with much of the current data (Chapter 4).

The rayleigh unit is useful in upper-atmosphere research but is awkward in engineering applications because of the dependence of the quantal energy on wavelength. Also, in the case of emissions that are not monochromatic (for example, the nightglow 'continuum'), it becomes necessary to introduce a wavelength interval; this is often accomplished by the use of the rayleigh per ångstrom, $R\ \text{Å}^{-1}$.

The factor 4π in the definition can be interpreted intuitively as sweeping a photometer completely around a centimeter column of upper-atmosphere emission and capturing all the quanta from the column. Thus, if the column's effective thickness is 10 km (or 10^6 cm), as is approximately true in the lower ionosphere, then, recalling the 10^6 in the definition, the rayleigh brightness of a particular emission becomes the number of quanta (or atomic transitions) per cm^3 per s, which brings the investigator into direct quantitative contact with the basic physics of the nightglow phenomenon.

In Table 2-VII, we show the interrelationships of three alternative brightness units. Actually, the expressions brightness and intensity, both of which appear in the literature of the light of the night sky, are poorly chosen – better choices would be luminance and radiance. For a table giving the interrelationships among the luminance units commonly used by laboratory workers in photometry, see Walsh (1958).

TABLE 2-VII

Conversion factors for units of surface brightness at 5300 Å

Unit of surface brightness	S_{10}(vis)	$R\ \text{Å}^{-1}$	B (erg cm^{-2} s^{-1} ster^{-1} Å^{-1})
1 S_{10}(vis) (luminance)	1	4.40×10^{-3}	1.31×10^{-9}
1 $R\ \text{Å}^{-1}$ (radiance)	227	1	2.98×10^{-7}
1 B (quantal output)	7.62×10^8	3.35×10^6	1

References

Bok, B. J. and Bok, P. F.: 1968, *The Milky Way* (3rd ed.), Sky Publishing Corp., Cambridge, Mass.

Herschel, Sir William: 1802, *Phil. Trans.* 495.

Hoskin, M. A.: 1963, *William Herschel and the Construction of the Heavens*, Oldbourne History of Science Library, Oldbourne, London.

Hunten, D. M., Roach, F. E., and Chamberlain, J. W.: 1956, *J. Atmospheric Terrest. Phys.* **8**, 345.

Olbers, W.: 1826, *Edinburgh New Phil. J.* **1**, 141.

Rayleigh IV, Robert John Strutt: 1930, *Proc. Roy. Soc., London A* **129**, 458.

Roach, F. E. and Megill, L. R.: 1961, *Astrophys. J.* **133**, 228.

Seares, F. H., van Rhijn, P. J., Joyner, M. C., and Richmond, M. L.: 1925, *Astrophys. J.* **62**, 320 (Mount Wilson Publ. 301).

Smith, L. L., Roach, F. E., and Owen, R. W.: 1970, AEC Research and Development Report BNWL-1419, UC-2.

van Rhijn, P. J.: 1925, Publications of the Groningen Astronomical Observatory, No. 43.

Walsh, J. W. T.: 1958, *Photometry* (3rd ed.), Dover Publications, Inc., p. 520, App. III, Table II.

Sir William Herschel, 1785; by L. F. Abbott.
(Courtesy, National Portrait Gallery, London)

PLATE II   SIR WILLIAM HERSCHEL
                                    1738–1822

Sir William Herschel, discoverer of the planet Uranus, was a versatile genius.

The first 20 years of his adult life were spent as a successful musician – organist, teacher, composer, conductor. During these years, however, he developed increasing interest in scientific matters – first through mathematics and its relationship to music theory, and then in astronomy and physics.

Although only an amateur astronomer in those years, he had carried out correspondence with recognized astronomers and written modest articles. Not until he discovered the new planet, Uranus, in 1781, was he accorded recognition as a professional, awarded the Copley Medal by the Royal Society, and elected a Fellow of the Society. Soon thereafter, he was appointed Astronomer Royal – the King's astronomer; this position gave him a permanent income and thus time and freedom to conduct research without having to depend on music for his livelihood. For the rest of his long life, William Herschel devoted his considerable talents to many facets of the field of astronomy.

Born in Hanover, Germany, on November 15, 1738, Frederick William was the second son of an oboist in the Royal Hanoverian Guards. Although uneducated by today's standards, the father encouraged his children in intellectual curiosity and love of nature. Having great interest in astronomy, he pointed out the features of the night sky to his children and discussed with them the work of such scientists as Kepler, Galileo, and Newton.

In 1755, William was sent to England with the Hanoverian Guards during the Seven Years' War. Although his health failed and he deserted to return to Germany, he was soon invited back to England to instruct a military band. His talents obtained him increasingly desirable situations until, in 1766, he became organist for the Octagon Chapel at Bath. In that locale, he became more and more interested in scientific matters – 'fluxions' (today's differential calculus), optics, physics, and astronomy. He developed considerable skill in devising optical instrumentation and refining current techniques, and devoted much of his night activities to observing the sky with instruments of his own making, one of which – the 40-ft-long reflector shown in the reproduction – was one of the largest telescopes of the time.

During this period, William began to be lonely for his family, so he invited his sister Carolina, 12 years younger than he and also unmarried, to join him in England, initially to help him as a singer in his musical activities. With his increasing devotion to studies of the night skies, however, she became adept at assisting him in that as well, recording his observations and so freeing him from the necessity of moving his eyes from their intent search for new and wonderful discoveries.

Forty-foot telescope designed, constructed, and used by Herschel at Slough.

Inasmuch as she had been working closely with William for several years before his discovery of the 'comet' that turned out to be Uranus, it is probable that she took the notes on the night that he made that significant observation. Her observational abilities developed sufficiently that, over the years, she discovered eight comets herself, with the 7-ft Newtonian sweeper that William gave her.

In 1788, when he was 50, William Herschel married a much younger English woman, by whom he had a son, John, who also became an eminent scientist. Although the marriage first estranged Carolina, William's wife was consistently friendly to her, and the birth of their son restored the close family bonds.

Although his discovery of Uranus was the feat that made him famous, William Herschel made other major contributions to the field of astronomy that are of equal, and possibly even greater scientific value – the refinement of telescopic optics; his analysis of the 'structure of the heavens', the forms of the Galaxy and nebulae, and a theory of the continuing process of their formation, from 'birth' through stages of development and growth to decline, old age, and 'death'; the period of rotation of Mars, to within 5 s of the value accepted today; his theory of self-luminous material, usually surrounding specific stars, as potential future stars; and numerous others, such as his early work on star counting.

Working with a 20-ft Newtonian reflector of 17.8-in. aperture, Herschel embarked on a long-range program of star-density estimation, which he termed 'star-gaging'. Directing his telescope to a particular part of the sky, he counted the stars visible in the field at each setting of the instrument, then moved to a nearby field and repeated the process. He initially assumed uniform distribution of the stars throughout space but, as he continued to compile data he abandoned this view, commenting in 1802, '... this immense starry aggregation is by no means uniform. The stars of which it is composed are very unequally scattered, and show evident marks of clustering together into many separate allotments' (*Phil. Trans.* 1802:495). This work, carried out over many years, became the basis of the star counts ably carried on and refined by later astronomers such as Kapteyn and Bok.

In this day of specialization in all fields, including astronomy, it is challenging to realize the tremendous scope, breadth, and sophistication of intellectual investigation this completely self-educated man was able to bring to bear on his chosen field of science.

William Herschel was truly a giant among astronomers – a first-magnitude star in his own Galaxy.

Professor J. C. Kapteyn, 1919:
photograph taken by F. Ellerman.
(*Courtesy, Mount Wilson Observatory*)

PLATE III   JACOBUS CORNELIUS KAPTEYN
1851–1922

It was not until more than a century after Herschel wrote '… this immense starry aggregation is by no means uniform' (*Phil. Trans.* 1802:495), that the Dutch astronomer Jacobus Cornelius Kapteyn devised an appreciably more refined technique for determining the numbers and distribution of stars in the heavens.

As one of fifteen children of the director of a boarding school in Barneveld, The Netherlands, Kapteyn's aptitude for science had been nurtured since early childhood. Not only Jacobus but several of his brothers became leaders in science. He obtained an excellent education at the University of Utrecht, where he received his doctoral degree in physics in 1875. The only position available at the time, however, was that of observer at the Leiden Observatory, and he accepted that. As a consequence, his scientific talents were bent toward astronomy – to physics' loss! Three years later, at age 27 – very young for the honor – he was named full professor in astronomy at the University of Groningen. Shortly thereafter he married Catharine Elizabeth Kalshovan, by whom he had two daughters and a son.

Despite the honor of the professorship, Kapteyn was frustrated because the University of Groningen had no observatory or observing instruments. Like Herschel, he was interested in the structure of the Universe (Galaxy) and to study that, he needed observing instruments. Fortunately, he secured permission to use the meridian circle of the Leiden Observatory during a professor's vacations. On this instrument, he planned a careful program for determining stellar parallaxes, introducing the differential method of observations in right ascension. To that time, the parallaxes of only 34 stars had been obtained, the most effective means being an instrument called a heliometer, and it was a slow technique.

Kapteyn's new approach yielded parallax information of accuracy comparable with that of the heliometer, and more rapidly, but was still not speedy enough for his requirements. By 1889 he had devised an ingenious photographic technique for measuring the parallaxes of large numbers of stars quickly and accurately.

The general problem of establishing the structure of the Galaxy was furthered by his speeded-up methods, but still was obscure because of lack of data. He soon realized two important points – first, by measuring proper motions of stars, he could determine their parallaxes with respect to the ever-increasing base line of the Sun's motion through space, and second, for the structure of the universe, it was not necessary to know the distances of individual stars, but rather the mean distances of groups of stars at different galactic latitudes.

Throughout his life, he adhered to a highly practical approach to scientific investigation. He could visualize the overall aspects of a major problem and, at the same time, develop techniques for utilizing limited available information and work out ways of obtaining additional and better data and refining the analysis and application of the new data to the problem. He focused his life's efforts on the galactic-structure problem, and this ability to expand and refine at the same time, iteratively, was an invaluable asset.

Kapteyn's investigations on proper motion led to his discovery of two star streams. Instead of random motion of stars, as had long been assumed, he found that the distribution of motion consistently tended in two preferential directions. Further, the lines seemed to converge on two points – one just south of α Orionis, the other just south of η Sagittarii. These findings convinced him that these two star streams were parallel to the lines joining our own solar system with the two apparent end points.

Such a concept necessarily influenced the then-current ideas about the structure of the Galaxy. But immense quantities of additional data, from far-flung sources, were needed to provide more reliable measurements and data for fainter stars.

At this point Kapteyn's genius for organizing the meticulous details necessary for a concerted attack on this major problem came to the rescue. He initiated steps to establish an international network of cooperative observing institutions. To systematize the monumental task of obtaining the desired data, in 1906 he published what has become known as his famous Plan of Selected Areas – 206 carefully selected regions over the entire celestial sphere – in which the observers would secure measurements on magnitude, proper motion, parallax, class of spectrum, and radial velocity for stars recorded on graded photographs centered on the selected regions.

In those days of limited, slow travel and communication, such a project might well have seemed impossible to many scientists. Kapteyn, however, being a gregarious man who had traveled widely, enjoyed the friendship as well as the professional respect of astronomers all around the world. This wide acquaintance made possible the tremendous coordinated activity of collecting data on his Selected Areas. By the time of his death, in 1922, the data that had poured in from all over the world had made possible the highly refined picture of the structure of our Galaxy that was essentially what we have used until very recently – more than 50 years after he conceived the project.

As in most areas of science, the solution of one problem brings up a host of new questions. And this has been true of Kapteyn's galactic structure – in the past decade radio astronomers have begun to find anomalies in the picture and have initiated new directions of investigation with their own new arsenal of research tools.

Kapteyn's contributions to our knowledge of the structure of the universe brought him well-deserved honors from all over the world.

Bart J. Bok with Telescope at Kitt Peak, Arizona.
(Courtesy, B. J. Bok)

PLATE IV   BART JAN BOK

1906–

Holland has contributed more than its share of outstanding astronomers in the past century, particularly in the area of galactic-structure research. In addition to Kapteyn, another Netherlander, Bart Jan Bok, has further refined the counting approach and contributed to a more precise shape-and-size description of our Galaxy than was possible before his time.

Born in Hoorn, The Netherlands, in April 1906, Bok studied at Leiden University. Although he studied there several years after Kapteyn's death, the influence of the Selected Areas proponent was still strong and undoubtedly served as inspiration to many of the young students, including Bok, who completed his studies there in 1927.

At the 1928 General Assembly of the International Astronomical Union, held in Leiden, Bok was on the Reception Committee, where he chanced to meet a young astronomer from Smith College, Priscilla Fairfield. After a year's courtship, he persuaded her to marry him, and they moved to Harvard University.

Harvard had a strong Astronomy Department, filled with eager and innovative graduate students and young professors, and Bok fitted in with ease, enjoying the stimulating atmosphere and associates. He obtained his American citizenship in 1938.

During his long career at Harvard – 1929 to 1957 – Bok held the G. R. Agassiz and R. W. Willson fellowships in the early years, became assistant professor in 1933, associate professor in 1939, and Robert Wheeler Willson professor from 1946 to 1957. He served as Associate Director of Harvard Observatory from 1946 to 1952.

Even in those busy years, he found time for research in his own areas of interest – galactic structure, in particular. One of his early researches, in the 1930s, was conducted on photographs of the constellation Carina in the southern sky – visible beyond the 'upper edge' of the Sagittarius arm of our Galaxy (see Figure 2-6). Some of the results of that study contributed to the development of his picture of the Galaxy as given in his book, *The Distribution of the Stars in Space*, published in 1937, which covered the state of knowledge at that time on the basis of the earlier work of Kapteyn, van Rhijn, and Seares, as well as of Bok and his students.

Astronomy students at Harvard in those early years counted millions of stars under Bok's tutelage, and he then placed many of them in strategically located departments of astronomy in the southern- and mid-U.S. universities. Each member of this 'star-counting brigade' was responsible for a different section of the Milky Way. The original purpose of the brigade was to define the structure of our Galaxy in the general vicinity of our solar system, but it soon outgrew that limitation. With the tremendous quantities of data generated, it became possible to fill in many of the gaps in the then very sketchy picture of our Galaxy. Gradually the emphasis shifted from studies of local stellar distribution to investigations relating to the spiral structure of our Galaxy. Additional inputs in the past decade, from such newcomers to the field as the radio astronomers, permit us to assume with relative certainty a rotating spiral structure with several arms, at least three of which have sections in Earth's quadrant of the Galaxy (Figure 2-6).

In the 1940s, Bok initiated concentrated investigation of globules, the smallest dark nebulae, obtaining results that led him to stress their importance as protostars. Later work, by both him and other researchers, has confirmed and expanded this concept, which has become important in studies of the life cycles of stars and star clusters.

In 1957, Bok accepted the position of Head of the Department of Astronomy at the Australian National University in Canberra and was also named Director of the Mount Stromlo Observatory. In Australia, he concentrated especially on research of the brilliant and beautiful Magellanic Clouds, appendages to the galactic structure, and on work relating to the spiral structure of the Carina Section of the Southern Milky Way.

He continued this line of research after returning to the United States, where he was appointed Head of the Department of Astronomy at the University of Arizona and Director of Steward Observatory in 1966. From that base, he was able to do considerable work on the spiral structure of the southern Milky Way using the telescopes of Cerro-Tololo Inter-American Observatory in Chile. Working in close collaboration with radio astronomers and theoretical workers in the field, Bok and his associates (principal among them, Priscilla F. Bok) have extended our knowledge of southern spiral structure in the Carina-Centaurus Section to distances up to 8000 pc from the Sun. Kinematically and structurally, the Carina Feature is now one of the best-known spiral features of our Galaxy.

Two key elements have characterized Bok's long career in astronomy. First, he has maintained meticulous use of statistical methods in studying the galactic structure and interstellar matter; this tenacious devotion to obtaining and analyzing massive quantities of data on which to base a concept – and of continuing to obtain and analyze data to confirm, refine, and expand the concept – has put him in the enviable position of being able to stand solidly on early statements and use them as foundations for future work. Second, his wide range of interests and lively curiosity about new questions that arise in solutions of old problems have kept him in the vanguard of 20th-century astronomy.

Since retiring from his official posts at the University of Arizona in 1970, Bart Bok and his wife Priscilla have completed the revision of their *The Milky Way* (1968; first published in 1938). He continues to publish articles on past research, and there are still those new questions that lure him into fresh research efforts.

Bart Bok has made, and continues to make, major contributions to astronomy generally, and particularly to the field of galactic structure, on which he has focused the spotlight during the greater part of his long professional life. And he is still going strong!

135 st Street

CHAPTER 3

THE ZODIACAL LIGHT AND GEGENSCHEIN

The projection of the Earth's orbit on the celestial sphere, the ecliptic, delineates the apparent annual path of the Sun against the star background. The other major members of the solar system – the planets and their moons – move within a few degrees of the ecliptic as a consequence of their intrinsic orbital motions combined with that of the Earth.

Prehistory does not, of course, document the first of our progenitors who first took advantage of his new, erect position to look inquisitively at the sky, noting first the Moon's steady march among the stars as it went through its monthly waxing and waning, then that other bright objects (the planets) wandered more slowly against the same general stellar background.

The ancients of different civilizations seem to have independently animated the ecliptic circle with 'living beings', which led to the Greek word *zōidiakos* and our present zodiac for the region along and near the ecliptic. The ancient lore of the zodiac included not only the imaginative outlines of the creatures of the 'zoo' into constellations (Figure 3-1) but also a whole library of tales of the activities, godly and ungodly, of the characters of mythology who traversed the zoological garden above the anxious eyes of the ancients and in turn served as their sometimes genial and sometimes stern chaperones.

That a relatively minor component of the solar system, a diffuse cloud of dust grains, also favors the ecliptic (zodiacal) region is evidenced by the existence of a photometric brightening toward the ecliptic due to the scattering of sunlight by the dust. The optical phenomenon carries the name of zodiacal light, and the physical that of the zodiacal cloud.

In the absence of moonlight and the aurora, the zodiacal light is the brightest component of the night sky. That it is not generally recognized is probably due to a number of factors. For example:

(1) It is best observed near the western horizon after twilight and near the eastern horizon before dawn, and is often confused with the twilight/dawn phenomena.

(2) Because it is brightest near the horizon, the potential zodiacal-light observer suffers from the competition of low-lying clouds and city lights.

(3) Since the ecliptic in mid and high latitudes is often significantly inclined toward the horizon, the zodiacal light is best seen in the tropics, where the ecliptic is more or less vertical with respect to the horizon.

For a well-located observer, the zodiacal light is a splendid spectacle (Frontispiece), appearing visually as a cone of light oriented along the ecliptic and brightest at the

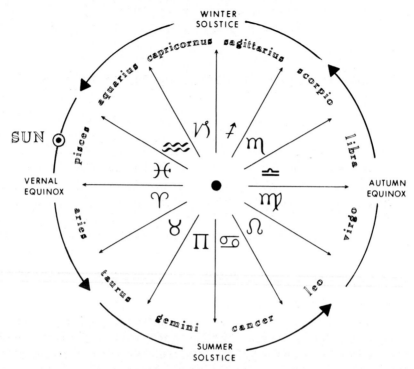

Fig. 3-1.   Circular plot of the Sun's annual traverse among the 12 constellations of the zodiac, in the direction of the arrows. The zero point for the measurement of right ascension and of ecliptic longitude is at the
intersection of the ecliptic and the celestial equator (the vernal equinox). In ancient times (Hipparchus, 2nd
century B.C.), this intersection occurred in the constellation Aries but, as a result of the precession of the
Earth's orbit, it now occurs in the preceding constellation, Pisces. It will move clockwise through one sign
each approximately 2000 years, arriving in Aquarius, for example, in about 4000 A.D.

horizon. It diminishes in both brightness and width as it extends well up toward the
zenith. Some observers have even reported visually detecting the zodiacal (ecliptic)
brightening along the entire ecliptic circle. At its brightest, as seen from the Earth's
surface, it is some three times as bright as the conspicuous southern Milky Way. At
its faintest, it is two to three times as bright as the integrated starlight at the galactic
pole.

Figure 3-2 shows photometric slices across the zodiacal light based on the tabulation in Appendix 3-A. At an elongation of 30° from the Sun, the width of the cone to
the visual observer is about 60° but, according to photometric measurements, the
zodiacal light extends out to the ecliptic pole, where its brightness – c. 78 S_{10}(vis) –
is a little less than that of the average integrated starlight – c. 105 S_{10}(vis) – and comparable with the nightglow continuum (see Chapter 4). This near equality of the three
components compounds the difficulty of their disentanglement. The apparent visual
width of the cone decreases at greater elongations, and the slices shown in Figure 3-2a
include most of the domain easily observable to the eye. Photometric measurements
confirm that there is an ecliptic enhancement at all elongation angles (Figure 3-2b).

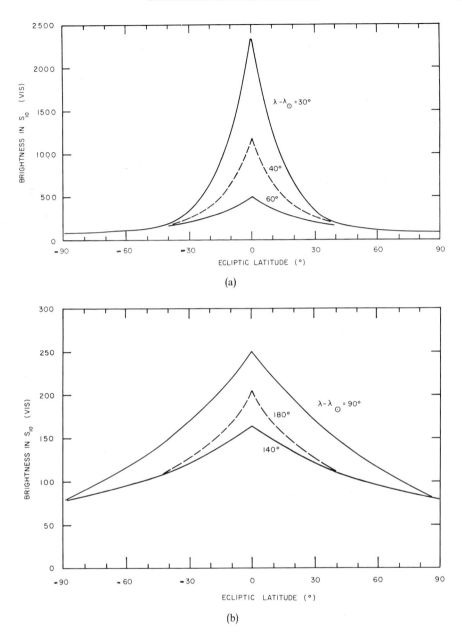

Fig. 3-2.   Photometric slices across the zodiacal light, based on the model in Appendix 3-A. Note that there is a 10-fold change of ordinate scale between Figure 3-2a (the bright region relatively near the Sun) and Figure 3-2b (the fainter region at high solar elongation angles). The slight increase in the brightness in the ecliptic from 140° to 180° is due to the gegenschein (see text).

The variation of zodiacal-light brightness along the ecliptic has been extensively studied. Figure 3-3 shows the brightness for that part of the phenomenon observable

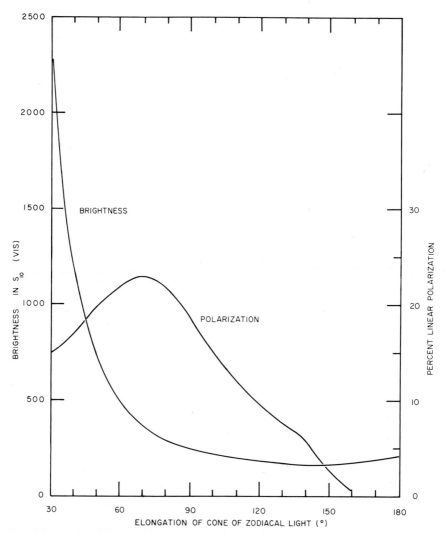

Fig. 3-3.   Plots of the brightness and linear polarization of the zodiacal light in the ecliptic for solar elongations of 30° to 180° (gegenschein). The brightness values are drawn from the tabulation in Appendix 3-A; the polarization values are from Weinberg (1963, 1964).

from Earth's surface* (elongation $\varepsilon > 30°$). The photometric gradient is very steep – from $\varepsilon = 30°$ to $\varepsilon = 45°$ along the ecliptic, the differential brightness is 1385 S_{10}(vis), almost the same as the difference at $\varepsilon = 30°$ from the ecliptic ($\beta = 0°$) to $\beta = 15°$, 1449 S_{10}(vis). These sharp photometric gradients give the visual impression of a definite edge to the phenomenon. Note in Figure 3-3 that the steep, exponential-type rise for the smaller elongations plotted suggests that the brightness of the zodiacal

* The zodiacal light can be observed without twilight contamination after astronomical twilight or before astronomical dawn, when the Sun is more than 18° below the horizon. Most observers have limited their measurements to elevations at least 10° above the horizon. Thus, if the ecliptic is vertical for the observer, the closest observation to the Sun is 28°.

light close to the Sun (angle-wise) must be very conspicuous. This point was brought out by van de Hulst (1947), who called attention to the fact that the terrestrial observations for $\varepsilon > 30°$ could be extended to the observations of the F (Frauenhofer) component of the solar corona on a log-brightness/log-elongation plot by a straight line with a slope of -2.4. Thus,

$$\log B = \text{const.} - 2.4 \log \varepsilon.$$

That there are two components of the solar corona was established by Grotrian (1934). One, labeled K, is the true corona, with streamer-like structures so often photographed at times of total solar eclipse. The other, the F corona, presumably due to sunlight scattered by dust in the line of sight, close to the Sun (in angle, but not physically), is less than 0.1 as bright as the K corona near the Sun's limb but dominates photometrically for elongations greater than one solar radius from the solar limb. It is this latter component of the corona that is associated with the zodiacal light.

In Figure 3-4 we have plotted the log of the zodiacal-light ecliptic brightness (from

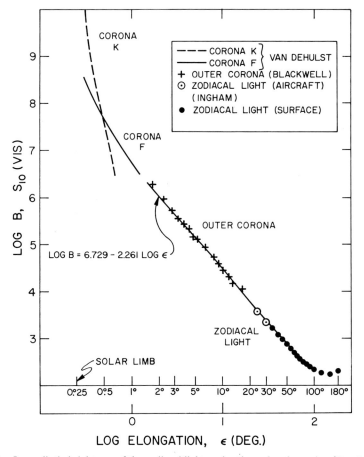

Fig. 3-4.   Log ecliptic brightness of the zodiacal light against log solar elongation (Roach, 1972).

Appendix 3-A) against log elongation angle. The slope of the dashed line connecting the faint end of the F corona (1°25 from the Sun's center) and the ground observation at $\lambda - \lambda_\odot = 30°$ is -2.37, in agreement with van de Hulst's early interpolation.

3.1. Polarization of the Zodiacal Light

The accurate quantitative measurement of the linear polarization of the zodiacal light is difficult because the direct observation is necessarily of the composite light of all the components of the light of the night sky, and in order to refer the measured polarization to the zodiacal light it is necessary to make arithmetic allowance for the dilution from the integrated starlight and airglow. In Figure 3-3 we include a plot of the linear polarization of the zodiacal light in the ecliptic plane according to measurements by Weinberg (1963). Over the elongation range 30° to 160° there is a well-defined maximum of 23% at 70°.

3.2. The Gegenschein

An extended faint glow* may be seen centered on that part of the ecliptic opposite the Sun. Its existence is clearly delineated in photometric measurements (see Figures 3-2b and 3-3) but it is a very elusive phenomenon to the visual observer. If there were no other sources of light than the zodiacal light and the gegenschein, the antisun region of the ecliptic would be about 20% brighter than the general surroundings. But the presence of integrated starlight and night airglow dilutes the contrast to about 10% under the best of circumstances. Successful observers have usually swept across the sky using averted vision, employing the off-axis part of the retina, which is especially sensitive if the eye is well dark-adapted (see Chapter 1, p. 3). Optimum observational conditions are obtained when: (1) the antisun direction is well away from the Milky Way (in February, March, April; September, October, and November), and (2) when the antisun direction is as high as possible above the horizon – that is, near local midnight.

   To some extent the difficulties involved in the visual perception of the gegenschein are reflected in its photography or in its accurate photometric measurement. Figure 3-5 shows two photographs of the same region of the astronomical sky made over a three-month span so that, in the one case, the antisun direction was included and in the second case it was some 90° distant. The presence of a glow more or less centered on the antisun direction in the first case is apparent. In the second photograph, the zodiacal band is clearly visible but the bright region of the first photograph is missing.

   Figure 3-6, based on the tabulation in Appendix 3-A, gives in isophote form the shape and extent of the phenomenon. It is seen to be oval in shape, extending some 40° from the antisun direction. In Figure 3-7 we show an isophotal map of the entire zodiacal-light/gegenschein phenomenon.

---

* The German word *gegenschein* has been internationalized to delineate the phenomenon: *gegen* (counter) *schein* (glow).

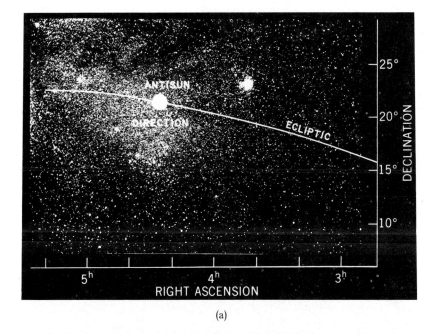

(a)

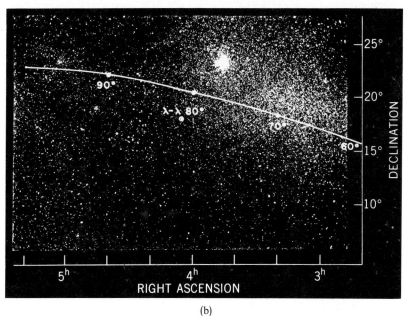

(b)

(Courtesy R. G. Roosen, McDonald Observatory)

Fig. 3-5.   Two photographs of the same astronomical region of the sky, taken three months apart. (a) Taken on December 1, 1966, this photograph shows the gegenschein as a general patch of light centered near the antisun direction. (b) Made on March 3, 1967, when the antisun direction was some 90° to the east, this photograph shows a part of the zodiacal band from elongations 60° to 80°; the gegenschein is, of course, no longer in the field of view.

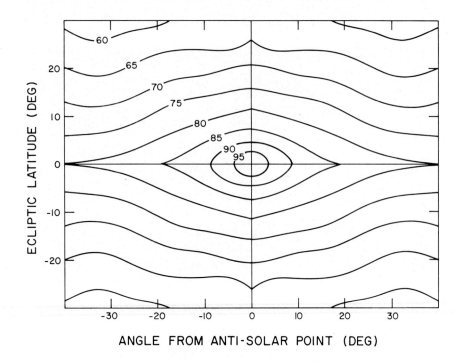

Fig. 3-6. Detailed plot of the photometric structure of the gegenschein (based on entries of Appendix 3-A), referred to relative brightness in percent of antisun value – 205 S_{10} (vis).

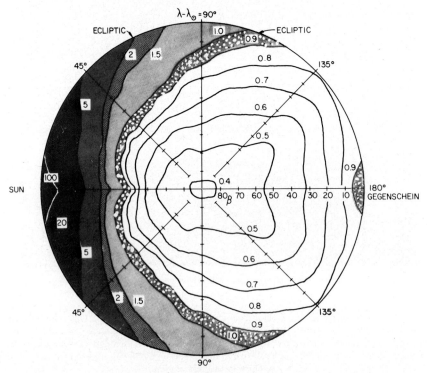

Fig. 3-7. Isophotal map of entire zodiacal light in polar-coordinate system (based on Appendix 3-A). Outer circle is ecliptic; center is ecliptic pole. Ecliptic latitudes (β) are indicated. The brightness unit is 1.0 for brightness in antisun (gegenschein) direction.

In Chapter 6, we shall return to the problem of the zodiacal light and the gegen-schein, with particular reference to the scattering of sunlight by interplanetary dust. We shall there discuss not only the general interplanetary dust cloud but also its possible perturbations in time and space.

References

Grotrian, W.: 1934, *Z. Astrophys.* **8**, 124.

Roach, F. E.: 1972, *Astron. J.* **77**, 887.

van de Hulst, H. C.: 1947, *Astrophys. J.* **105**, 471.

Weinberg, J. L.: 1963, *Photoelectric Polarimetry of the Zodiacal Light at* λ*5300*. Ph. D. Thesis, Univ. of Colorado.

Weinberg, J. L.: 1964, *Ann. Astrophys.* **27**, 718.

APPENDIX 3-A

The zodiacal light and gegenschein

Numbers in ecliptic ($\beta = 0$) taken from Figure 3-4. Off-ecliptic numbers (Roach, 1972)

Values of brightness in zodiacal light and gegenschein (S_{10}(vis) units)

$\lambda - \lambda_\odot$	90	85	80	75	70	65	60	55	50	45	40	35	30	25	20	15	10	5	$\beta = 0$	$\lambda - \lambda_g$
180	78	80	82	84	87	89	93	97	101	106	112	119	126	135	145	156	168	182	205	0
175	78	80	83	86	90	94	98	102	106	110	113	117	125	130	144	154	165	176	191	5
170	78	80	83	86	90	93	97	101	104	109	114	119	124	132	140	152	162	173	183	10
165	78	80	83	86	90	93	97	101	103	108	113	118	124	132	139	150	158	169	178	15
160	78	80	83	86	89	92	95	99	101	105	109	115	121	129	138	147	154	165	173	20
155	78	80	83	86	88	91	94	97	100	103	106	113	120	128	136	141	151	161	169	25
150	78	80	83	85	88	90	93	96	99	102	105	111	117	124	131	139	149	158	167	30
145	78	80	83	85	88	91	93	96	99	102	105	112	118	125	132	139	147	156	165	35
140	78	81	83	86	89	92	95	99	102	106	109	115	121	127	133	140	147	155	164	40
135	78	81	85	88	92	96	100	104	109	113	118	123	128	133	139	145	151	157	165	45
130	78	82	85	89	93	97	102	106	111	116	121	127	133	139	145	151	158	166	173	50
125	78	82	85	89	94	99	102	107	112	117	123	129	135	141	148	154	162	169	177	55
120	78	82	86	90	95	100	105	110	116	122	128	134	141	148	156	164	172	181	190	60
115	78	82	86	91	95	100	105	111	116	122	129	135	142	149	157	165	173	182	191	65
110	78	82	87	91	96	101	107	112	119	125	132	139	146	154	162	171	180	190	200	70
105	78	83	87	92	97	103	109	115	121	128	135	143	151	160	169	178	188	199	210	75
100	78	83	88	93	98	104	110	117	124	131	139	147	156	165	175	185	196	208	220	80
95	78	83	88	94	100	106	113	120	127	135	144	153	163	173	184	196	208	221	235	85
90	78	83	89	95	101	108	115	123	131	140	149	159	170	181	193	206	220	234	250	90
85	78	81	85	91	95	103	111	118	127	138	148	159	173	186	200	214	230	247	265	95
80	78	83	87	92	101	108	116	123	132	141	151	162	174	188	205	225	248	268	290	100
75	78	83	88	93	98	104	113	123	135	147	159	175	190	207	225	245	269	294	320	105
70	78	83	88	94	100	106	114	125	138	152	168	185	205	225	249	275	302	336	370	110
65	78	81	84	87	94	100	109	118	131	145	162	183	205	233	262	295	332	376	425	115
60	78	81	85	89	95	100	109	119	132	148	168	189	214	238	269	309	361	430	500	120
55	78	81	85	91	98	106	117	131	145	162	182	204	224	251	284	337	406	502	600	125
50	78	81	86	91	98	107	118	132	147	163	180	202	230	269	320	387	479	603	745	130
45	78	81	85	91	98	106	119	133	150	163	185	209	241	288	351	427	552	714	945	135
40	78	83	88	93	100	107	118	132	148	166	190	216	263	324	406	507	655	871	1185	140
35	78	80	85	88	94	100	109	120	136	152	178	219	288	380	504	676	910	1200	1595	145
30	78	81	84	89	95	101	112	126	143	164	191	240	331	457	635	881	1230	1710	2330	150
25	78	81	86	89	96	103	114	131	145	170	195	260	390	500	740	1050	1445	2190	3700	155
20	78	81	86	89	96	102	115	132	148	175	200	280	430	600	850	1260	1700	2880	6100	160
15	78	81	86	89	96	102	115	134	151	180	210	295	470	710	950	1580	2190	3800	12000	165
10	78	81	86	89	96	101	115	137	158	190	230	320	520	810	1100	2000	2950	5890	29000	170
5	78	81	86	89	96	100	116	143	166	205	250	350	575	980	1350	2880	4470	10700	140000	175
0	78	81	86	89	96	100	119	169	200	250	320	400	630	1020	1820	6170	15800	89100	7.59(8)	180

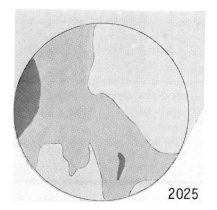

CHAPTER 4

THE NIGHT AIRGLOW OR NIGHTGLOW

Our wandering planet is sheathed in a nocturnal glow that usually goes unnoticed. The glow is due to a variety of physical causes and, although it is often referred to as the night airglow or nightglow, it is probably better to speak of a number of nightglows, not only because of their numerous physical origins but also because of the multiplicity of their temporal and geographical changes.

Physical causes of the glow include chemical reactions of the neutral constituents of the upper atmosphere resulting in the emission of light; reactions involving ionized constituents also resulting in light emission; excitation of atmospheric constituents not in the Earth's shadow by sunlight; and excitation by incoming energetic charged particles guided along geomagnetic lines of force. The first two sources of the glow are omnipresent and contribute to the light of the night sky for all Earth-bound observers. The third has been observed recently from rockets and satellites, but much of its radiation is in the UV and does not penetrate to the Earth's surface. The fourth results in the polar aurora, which tends to occur in zones some $20°$ or $25°$ from the geomagnetic poles; the occurrence of the polar aurora has a strong dependence on the sunspot cycle.

The nightglow is a constant Earth 'envelope' – every astronaut has seen it on every 'night' orbit of the Earth, as a luminous annulus following the curved Earth surface (Figure 4-1). An astronaut in orbit around the Earth is particularly well located to see the nightglow, since his tangential view through the emission layer increases the effective path length some 30-fold relative to a zenith observation on the Earth's surface. The correspondingly enhanced brightening of the emission layer near 100 km puts it well above visual threshold.

From the Earth's surface there is a zenith-to-horizon increase in the radiance of the nightglow (Appendix 4-A), which can be observed in spite of the photometric dilution of the effect by the other constituents of the LONS. As a matter of fact, *each*

• α HYDRA

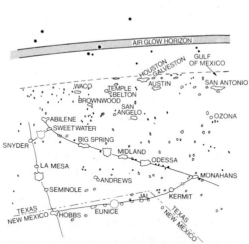

Fig. 4-1.   Photograph and matching map of nightglow made during the Aerobee rocket flight of November 30, 1964. The nightglow is seen as a sort of annulus above the moonlit Earth over a large portion of Texas, with highways and towns identified on the map below (from Hennes and Dunkelman, 1966: 756–757).

major component of the LONS tends to dilute the contrasting effects of its competitors. It has been known since the star gaging of William Herschel that the region of the Milky Way is some 10 times as rich in stars as the sky well away from it (see Chapter 2). But, if the reader will position himself on a moonless night where he can look

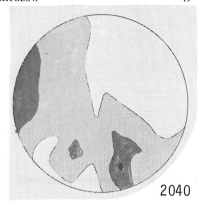

2040

at the sky*, he will quickly convince himself that the Milky Way – even in its brightest parts, such as the Sagittarius-Scorpius region – does not *look* 10 times as bright as the non-Milky Way sky. Obviously there is a dilution of the contrast by light sources of nonstellar origin. We have seen in Chapter 3 that a significant part of the photometric dilution is due to the presence of the zodiacal light, but this effect can be minimized if the observation is made during the middle of the night, when neither the bright evening nor the bright morning zodiacal light is a marked feature of the sky. If he will then make a vertical scan from zenith to horizon – avoiding the Milky Way as much as possible, and also avoiding the general east and west horizons because of possible residual zodiacal light – he can usually note that the night sky increases perceptibly in brightness toward the horizon as a result of the nightglow.

An early application of the zenith-to-horizon scan method was made by Yntema (1909), who dubbed the terrestrial component of the LONS the 'earthlight'. This component was investigated by Lord Rayleigh IV – the 'airglow Rayleigh' (see Plate V) – in the 1920s in early efforts to distinguish among various sources of the LONS. Having determined that this faint component was essentially unpolarized, he was able to eliminate the polarized zodiacal light as its major source. He then initiated photometric studies over the spectrum, obtaining evidence that led him to conclude that the faint light was separate from the polar aurora, and he named it the 'nonpolar aurora'.

The term 'airglow' was suggested by Otto Struve** to C. T. Elvey, who introduced it into the literature as '... the light emitted by the upper atmosphere other than that which is known as the polar aurora'[†] (1950). The generic term 'airglow' is often

---

* We assume that our observer has become reasonably dark-adapted; he might carry with him a black card to hold up against the sky – the contrast in brightness between the card and the sky is convincing evidence that the night sky is truly 'bright'.

** Though not an active worker in airglow research, Struve, as director of the Yerkes/McDonald Observatories, encouraged some of his junior associates (including F. E. Roach) to include geophysical airglow studies in general investigations of the light of the night sky.

[†] Elvey's definition carries a reasonable meaning to active workers in the field, even though it is often not possible to be definitive about the presence or absence of a polar aurora at any particular time or place. A random observation must of necessity await the *ex post facto* determination of auroral activity before it is known whether airglow has been observed.

particularized for specific studies – for example, day airglow (or dayglow), twilight airglow, night airglow (nightglow).

4.1. The Nightglow – Static or Dynamic?

The comparative difficulty in 'seeing' the nightglow has led to its being described as a quiescent, well-behaved phenomenon. Recent photometric studies indicate otherwise, however, and to counter that static view, we illustrate the dynamics of the tropical nightglow by a series of circle maps of 6300 Å centered on the Haleakala station in Hawaii. The 18 maps have been located on the upper right corners of the recto pages of this chapter in such a way that, by flipping the pages, the reader can produce a cinema effect as the localized patches of radiation wax and wane. In this time-lapse representation, the natural phenomenon is speeded up about 1000-fold – that is, an entire night of 6300 Å patterns is scanned in a few seconds. If the tropical enhancements, typically lasting an hour, were visible in real time, they would probably seem a bit sluggish; but to the photometric observer of the nightglow, they are quite dramatic, often requiring frequent changes of the sensitivity setting of the photometer to keep the deflections on scale.

As you will note on examining the corners individually, the time-lapse intervals extend over a period of 5 hr 30 min, during which more than a single waxing-waning cycle is revealed. If this phenomenon appeared in 'living color' in the night sky, it would be only slightly less spectacular than the polar aurora. As it actually occurs in the red below visual threshold, we regretfully can 'see' it only photometrically. Nevertheless, as the cinema effect demonstrates, it is a decidedly dynamic phenomenon.

4.2. Sources of Nightglow

The three principal components of the LONS (starlight, zodiacal light, and nightglow) – none of which is of overwhelming relative brightness – must be disentangled by successive approximations. The discrimination is aided by the fact that the nightglow, in contrast to the zodiacal light and the integrated starlight, includes many spectroscopically discrete emissions that can be resolved by spectroscopic methods – either a spectrograph of modest dispersion or filters with transmission bands centered on the wavelength of a particular radiation. For example, it was early established that the green line due to atomic oxygen (O I) at 5577 Å, a prominent auroral emission, (1) is always present in the night-sky light, (2) tends to increase in brightness toward the horizon, and (3) is not appreciably influenced by geomagnetic activity. Rayleigh IV made the first reliable measurement of the absolute brightness of this line and defined the photometric unit for expressing that measurement quantitatively. The unit, officially named the *rayleigh* (R)* in his honor in 1956, is in common use in airglow research today.

Both the nightglow and the aurora contribute to the LONS of an Earth-surface
_____
* The unit is defined in Appendix 2-B.

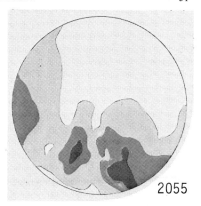

2055

observer. The excitation of an aurora is the result of the transfer of some of the kinetic energy of the motions of impinging particles to the resident atoms or molecules in the upper-atmosphere zone. All other nocturnal upper-atmosphere radiations are night-glow. Because of its concentration in reasonably well-defined geographic zones*, we propose to treat the polar aurora as a perturbation – albeit a dramatic one – on the general planetary nightglow situation and discuss it separately at the end of this chapter.

There has been some confusion over the semantics of the expressions *nightglow* and *aurora* resulting in part from the contrast between the apparent dynamics of the two phenomena and in part from the original definition of the (night) airglow as upper atmosphere emissions of nonauroral origin. To some extent, a definitive appellation for a particular upper-atmosphere phenomenon producing radiation is not possible and, in such cases, it is probably wise to describe the phenomenon and search for its physical origins without undue preoccupation with names. A case in point is the M/SAR arc, which bears the earmark of an aurora in its undoubted orientation along geomagnetic parallels but is atypical in its very low excitation level, as discussed later in this chapter.

4.3. Photochemical Reactions in the Upper Atmosphere

In Figure 4-2 we show a schematic representation of the two-way interactions between light and chemical (or physical) changes in the upper atmosphere. Radiation (sunlight) produces changes in the atmospheric components by dissociating the molecular species into their atomic constituents and by removing electrons from neutral atoms or molecules and producing a plasma of charged particles intermingled with the still-predominant neutral particles. The products of dissociation or ioniza-tion interact to produce radiation (nightglow) as they recombine in the course of their small-scale (kinetic) or large-scale (geostrophic) motions.

In Tables 4-I and 4-II we list the principal nightglow radiations in decreasing order' of absolute intensity. Table 4-I includes emissions resulting from the photo-

---

* Although an aurora may occur anywhere above the Earth's surface, it is statistically concentrated in the so-called auroral zones, small circles about 23 deg from the geomagnetic poles.

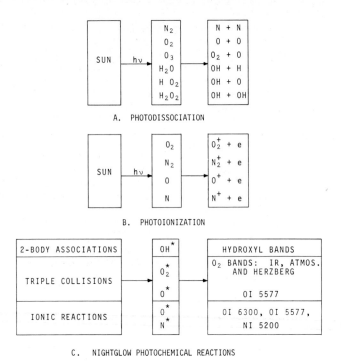

A. PHOTODISSOCIATION

B. PHOTOIONIZATION

C. NIGHTGLOW PHOTOCHEMICAL REACTIONS

Fig. 4-2. Photoreactions in the upper atmosphere.

chemical association of neutral-atmosphere components, Table 4-II from reactions in the ionosphere involving ionized components. These groupings bring out the fact that the neutral reactions are effective in the 90 to 100 km region, the ionic reactions between 250 and 300 km.

4.3.1. Photochemical Association

The 90 to 100 km region has often been called the 'chemical kitchen' because of the large number and the complexity of the chemical reactions that occur there. The complexity arises because the composition of the atmosphere in this region is significantly altered by the UV solar radiation. The total number density is shown in Figure 4-3. The principal effect is the dissociation of the O_2 molecule into two O atoms by solar radiation of wavelength shorter than 1750 Å, corresponding to the dissociation potential (7.049 eV) of the O_2 molecule (Figure 4-4). Superimposed on this photo effect are gravitational and dynamical influences such as the differential diffusion of components of different molecular weights and the somewhat compensating homogenization resulting from convection and turbulence.

In Figure 4-5 we show a model of the distribution with height in the 65 to 105 km region of some of the atmospheric components of interest in interpreting the nightglow. Atomic O (at midnight) peaks at about 92 km. Molecular and atomic oxygen are, according to this model, of equal concentration at 100 km. Molecular nitrogen (N_2) is not significantly dissociated at this altitude – its plot parallels the 'total' curve over this domain.

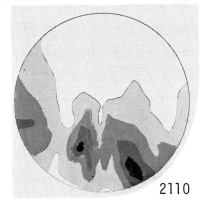

2110

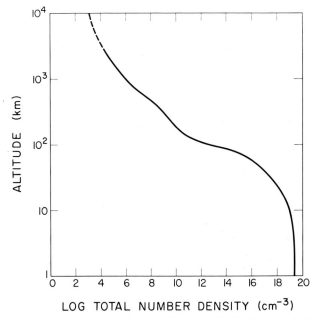

Fig. 4-3.   Number density in the atmosphere as a function of altitude above the Earth's surface.

A comparison of the absolute quantal intensities in Table 4-I and the concentrations in Figure 4-5 reveals an apparent paradox – the overwhelmingly strongest emitter, hydroxyl (OH), is due to one of the very minor constituents, which is five or six orders of magnitude less abundant than the major atmospheric components – O, O_2, and N_2. To take a numerical example, the ratio of the peak concentrations of OH and O is about $10^7/10^{12} = 10^{-5}$. The ratio of quantal output of the OH system of bands to that of the O I line at 5577 Å is $4\,500\,000/250 = 18\,000$. Thus, if the emission is due to some mechanism concerned with the direct excitation of the emitting species, then the relative excitation efficiencies are

$$OH/O = 18\,000/10^{-5} = 1.8 \times 10^9.$$

TABLE 4-I

Nightglow radiations due to excitation by chemical association ordered by emission intensity

Emitter	Wavelength (Å)	Transition		E.P. (eV)	Probable reaction	Emission height (km)	Absolute intensity (R)	Remarks
		Upper level	Lower level					
Hydroxyl (OH)	3817–44702	Rotation-vibration bands ($v \leqslant 9$); $\Delta v = 1, 2 \ldots 9$		3.23 ($v=9$)	$H + O_3 \rightarrow OH + O_2 + 3.34$ eV	90	4 500 000	Strongest bands are in near IR (Figure 4-4)
Molecular oxygen (O_2)	12 700 (0-0) 15 800 (0-1) 19 000 (0-2)	$a\,^1\Delta_g$	$X\,^3\Sigma_g^-$	0.98	$O + O + M \rightarrow O_2\,(a\,^1\Delta_g) + M$ $k = 5.4 \times 10^{-34}$	90	80000	IR atmospheric bands
	7619 (0-0) 8645 (0-1)	$^1\Sigma_g^+$	$X\,^3\Sigma_g^-$	1.60	$O + O + M \rightarrow O_2 + M + 5.17$ eV	~80	6000	Atmospheric bands
	2600–3800	$A\,^3\Sigma_u^+$	$X\,^3\Sigma_g^-$	4.3	$O + O + M \rightarrow O_2 + M + 5.17$ eV	90	600	Herzberg bands
NO_2	5000–6500				$O + NO \rightarrow NO_2 + 3.1$ eV	90	250	Nightglow continuum
Atomic oxygen (O)	5577 2972	^1S ^1S	^1D ^3P	4.17 4.17	$O + O + O \rightarrow O_2 + O + 5.17$ eV $O + O + O \rightarrow O_2 + O + 5.17$ eV	90 90	250	Observable only from rockets or satellites above ozone layer
Atomic sodium (Na)	5890, 5896	P	^2S	2.09		~92	50	Strong seasonal variation

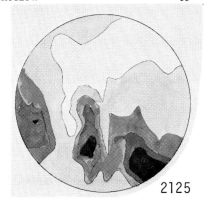

2125

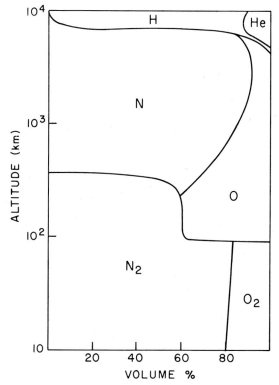

Fig. 4-4. Schematic representation of changes in chemical composition of the atmosphere with altitude. Note the abrupt change near 100 km, where O_2 molecules are significantly dissociated into O atoms.

It is obvious that we must anticipate entirely different types of excitation mechanisms for the two cases.

4.3.1.1. *Hydroxyl (OH) Emissions*

We have commented on the quantal output of OH emissions relative to that of the

TABLE 4-II

Nightglow radiations due to excitation by ionic reactions

Emitter	Wavelength (Å)	Transition		E.P. (eV)	Probable reaction	Emission height (km)	Absolute intensity (R)	Remarks
		Upper Level	Lower Level					
Atomic oxygen (O)	1304	^3S	^3P	9.52	$O^+ + e \rightarrow O^* + h\nu$	250	150	Observed from satellites
	1356	^5S	^3P	9.14	$O^* \rightarrow O + h\nu$			
	6300	^1D	^3P	1.96	$O^+ + O_2 \rightarrow O_2^+ + O$ $k = 1.5 \times 10^{-11}$ cm^3 s^{-1}	300	100	Sporadic enhancements in the tropics associated with ionospheric disturbances
	6364	^1D	^3P	1.96	$O_2^+ + e \rightarrow O + O + 6.96$ eV $k = 1.9 \times 10^{-7}$ cm^3 s^{-1}			
	5577	^1S	^1D	4.17	$O^+ + O_2 \rightarrow O_2^+ + O$ $k = 1.5 \times 10^{-11}$ cm^3 s^{-1} $O_2^+ + e \rightarrow O + O + 6.96$ eV $k = 2.1 \times 10^{-8}$ cm^3 s^{-1}	300	20	This high-atmospheric component of 5577 is observed chiefly in the tropics during enhancement of O I 6300, 6364
Atomic nitrogen (N)	5198	$^2D^0$	$^4S^0$	2.37	$NO^+ + e \rightarrow N + O + 2.74$ eV $k = 4.1 \times 10^{-7}$ cm^3 s^{-1}	258	1	
	5201							

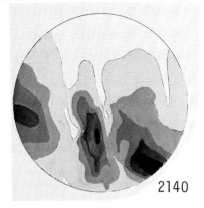

2140

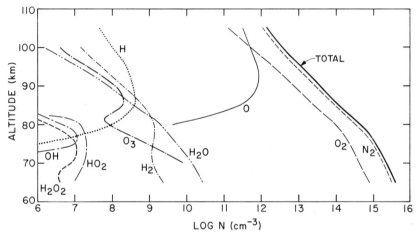

Fig. 4-5.   Graphical representation of variations of atmospheric constituents in the 65 to 105-km region at
midnight.

green line (5577 Å) of O I. The complicated system of OH rotation-vibration bands is
concentrated in the invisible near-IR (Figure 4-6). If all the OH radiation were concentrated in the visible part of the spectrum, the night sky would glow like a bright
aurora or midtwilight. It has been said that the history of astronomy would have
been seriously affected by such a contingency, since only the brightest stars and planets
would have been visible. The Milky Way, lost in the competing glow, would have been
discovered only as a small photometric 'noise' fighting for recognition with respect
to a very large OH 'signal'.

The OH system of bands is the result of transitions among the nine lowest levels,
v, of the lowest electronic state ($^2\Pi$) of the radical. No selection rules apply and all
combinations of Δv are permitted – 45 bands in all. Each band has a complicated
structure (Chamberlain, 1961:368, table 9.1; 556, table 13.1), such that the individual
lines spread out over some scores of Ångstroms. It is difficult to avoid some contamination by OH radiations in the case of observations of other nightglow emissions

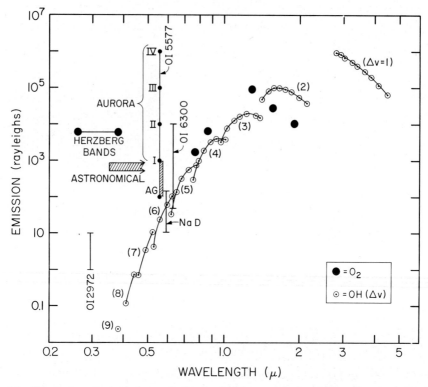

Fig. 4-6.   Principal upper-atmosphere emissions on emission vs. wavelength plot. Compare with Tables
4-I and 4-II.

with wide filters. For example, the 8-2 OH band surrounds the D_1 and D_2 lines of
atomic sodium (Na); the 7-1 band is near the O nightglow 5577 Å; and the 9-3 band
is near O I 6300 Å.

In the nightglow OH a prime observational fact is that bands with upper vibrational
level up to $v = 9$ (3.23 eV) have been observed, but none from level $v = 10$ (3.49 eV) or
higher. This restriction was noted early and led to the suggestion that the reaction
responsible for the emission is

$$H + O_3 \rightarrow OH + O_2 + 3.34 \text{ eV},$$

which is seen to be sufficiently exothermic to excite $v' = 9$ but not $v' = 10$. It remains
to inquire whether this reaction between two very minor atmospheric constituents
is sufficiently rapid to produce the observed quantal output. The rate coefficient,* k,

---

* The reaction rate constant k arises from the experimental law that the rate of reaction is proportional to
the product of the concentrations of the reacting substances. k is the constant of proportionality. Thus,
given a reaction $A + B + C = ABC$, the reaction rate is often written as a rate of loss, for example A, as
$d[A]/dt = k[A][B][C]$. This definition applies to all types of reactions, whether the left side consists of
one substance two, three, or more. The dimensions of k are different for different orders of reaction. Thus
s^{-1} for first order, $cm^3 s^{-1}$ for second order, $cm^6 s^{-1}$ for third order.

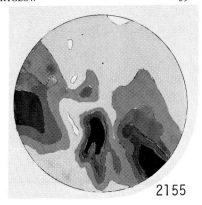

2155

of the reaction has been measured in the laboratory to be

$$k = 1.5 \times 10^{-12} \sqrt{T} \ \text{cm}^3 \ \text{s}^{-1} = 2.1 \times 10^{-11} \ \text{cm}^3 \ \text{s}^{-1}$$

for $T = 200\,\text{K}$. At a height of 90 km, the predicted quantal output using this rate coefficient is

$$Q = [\text{H}] \, [\text{O}_3] \, 2.1 \times 10^{-11} = 10^8 \times 10^8 \times 2.1 \times 10^{-11} = 2 \times 10^5$$

quanta $\text{cm}^{-3} \ \text{s}^{-1}$. At 85 km the prediction is 1.6×10^6 quanta $\text{cm}^{-3} \ \text{s}^{-1}$. For an effective layer thickness of 10 km (10^6 cm) the columnar emission is 1.6×10^{12} quanta, or 1.6×10^6 R. Thus the general order of the prediction is in agreement with the observations.

4.3.1.2. *Molecular and Atomic Oxygen Emissions*

4.3.1.2.1. *Role of triple collisions in nightglow emissions.* We treat the three O_2 bands (Table 4-I) and the O I lines 5577 Å and 2972 Å together because the present evidence suggests a single basic excitation mechanism. In 1931, Chapman called attention to the reservoir of energy stored in the upper atmosphere from the recombining of atomic into molecular oxygen by way of a triple collision in which the energy of recombination (5.17 eV) is available either as kinetic energy of one of the reactants (thus raising the temperature) or as potential energy to excite one of the reactants to a state from which emission occurs. Triple collisions are infrequent as compared with double collisions, but have the property of using some of the excess energy (5.17 eV in the case under discussion) for excitation even when energy resonance is not involved.* The general reaction proposed by Chapman is

$$\text{O} + \text{O} + M \rightarrow \text{O}_2 + M + 5.17 \ \text{eV}, \tag{4.1}$$

where M is any third body, atomic or molecular. Particular reactions pertinent to

---

* Resonance, on the other hand, is favorable in the case of double-collision excitation.

nightglow emissions are:

$$O(^3P)+O(^3P)+M \rightarrow O_2(a\ ^1\Delta_g)+M+4.19 \text{ eV} \tag{4.2}$$
$$O(^3P)+O(^3P)+M \rightarrow O_2(^1\Sigma_g^+)+M+3.57 \text{ eV} \tag{4.3}$$
$$O(^3P)+O(^3P)+M \rightarrow O_2(A\ ^3\Sigma_u^+)+M+0.87 \text{ eV} \tag{4.4}$$
$$O(^3P)+O(^3P)+O(^3P) \rightarrow O_2+O(^1S)+1.00 \text{ eV}. \tag{4.5}$$

Reaction (4.2) has a rate coefficient of 5.4×10^{-34} cm^6 s^{-1}. If the concentration of O is 2.8×10^{11} cm^{-3} and if M is N$_2$ with a concentration of 7.6×10^{13} cm^{-3}, we have for the quantal output, Q,

$$Q=6 \times 10^{36} \times 5.4 \times 10^{-34} = 3000 \text{ excitations cm}^{-3} \text{ s}^{-1},$$

or 3000 R for a column of 1 km (10^6 cm) thickness if each excitation results in an emission (no loss by collisional de-excitation). Thus the prediction is lower than observed for the IR atmospheric bands (80 000 R). It has been suggested that the atmospheric model (Figure 4-5) provides too low a concentration of O in the 90 km region, possibly because of the downward movement of O during the night from the 100 km region. For example, if O has a concentration of 10^{12} cm^{-3}, reaction (4.2) predicts an emission (with no collisional de-excitation) of 40 000 R for the IR atmospheric bands. However, the long mean lifetime of the upper state (3600 s) makes it subject to significant collisional de-excitation and imposes a severe restriction on the suggested reaction.

Reaction (4.5) has historical priority. Chapman introduced it as an excitation mechanism to produce O atoms in the ^1S state at a time when the 5577 Å ($^1D \leftarrow {}^1S$) line of O I was the best-known and most-extensively studied nightglow radiation. The reaction is the 'first born' of a large family. If the rate coefficient is the same as that for reaction (4.2), 5.4×10^{-34} cm^6 s^{-1}, then the peak quantal output is 540 quanta cm^{-3} s^{-1}, which corresponds to 540 R for an emitting layer of 10 km effective thickness (again assuming zero collisional de-excitation).

4.3.1.2.2. *Collisional de-excitation.* More than 40 reactions in the 100 km region have been listed among the allotropic forms of O and H: O$_3$, O$_2$, O, H$_2$, H, H$_2$O, HO$_2$, H$_2$O$_2$, OH. Of these reactions, some are directly concerned with the production of excited atoms or molecules that produce emissions. All the reactions are pertinent in an understanding of the total picture and, in particular, some of the reactions are competitive to emissions as de-excitation or quenching mechanisms.

Many of the nightglow emissions observed from the 100 km region are due to forbidden transitions. The upper excited atomic levels are metastable, having lifetimes long compared with the intercollision periods (see Table 4-I). The effect of collisional de-excitation during the long life of the typical metastable levels is now discussed for the case of the O I emissions 5577 Å ($^1D \leftarrow {}^1S$) and 6300 Å ($^3P \leftarrow {}^1D$). The energy-level diagram for the lowest states of O I (Figure 4-7) illustrates that the ^1S state has a mean life of 0.74 s and the ^1D state, 110 s. The collisional frequency at 90 km is

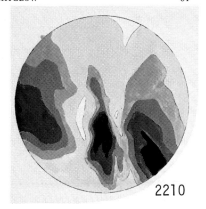

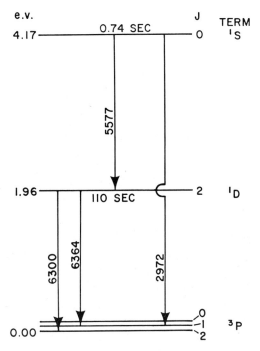

Fig. 4-7. Energy-level diagram of three lowest states of atomic O. All indicated radiations are due to transitions from long-lived (metastable) upper levels.

1.87×10^4 s^{-1}. Thus, during the time that an oxygen atom is procrastinating in the metastable ^1S state, it suffers 13 800 collisions and in the ^1D state, 2×10^6. Fortunately, the efficiency of de-excitation is low and only a small fraction of the collisions removes the atom from its excited state.

Figure 4-8 illustrates the change of de-excitation (quenching) with height based on an assumed quenching rate by collisions of 10^{-13} cm^3 s^{-1}. The deduced quenching is equal to the emission for the O I (^1S) state at a height of 100 km and for the O I (^1D) state at a height of 140 km. This is consistent with the facts that the 5577 Å emission

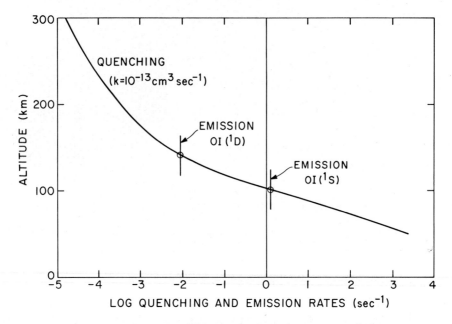

Fig. 4-8.   Collisional de-excitation (quenching) rate in the upper atmosphere as a function of altitude for assumed de-excitation coefficient of 10^{-13} cm^3 s^{-1}. Emission rates (reciprocals of the mean lifetimes) for atomic O states ^1D and ^1S are indicated.

$(^1D \leftarrow {}^1S)$ is observed at a height of 90 to 100 km and the 6300 Å $(^3P \leftarrow {}^1D)$ only at much greater heights.

4.3.1.3. *The Sodium (Na) D Lines in the Nightglow*

The yellow *D* lines of Na are ubiquitous in nature: They occur as absorption lines in stellar spectra in general and in the solar spectrum in particular; they show up as a component of the spectrum of interstellar space; they are scattered in the atmosphere from urban street lights; they plague the spectrochemist if he is not meticulously clean in his laboratory procedures; they appear in the twilight glow as the Sun illuminates the upper atmosphere; and, not surprisingly, they are found in the nightglow.

The very short lifetime of the upper atomic state, about 10^{-8} s, is one reason for the pervasive nature of the radiation. Such short residencies in the upper state mean that the $^2P^0$ Na atom at 90 km is able to radiate 5350 times between elastic collisions and about 16×10^6 times between quenching collisions. Even at sea-level pressures, there are about 50 atomic emissions per quenching collision.

The abundance of Na atoms in the upper atmosphere is about 100 cm^{-3}, too small to have been plotted on Figure 4-5. The low concentration is obviously compensated by the high transition rate in producing an observable radiation in the nightglow. An interesting feature of the Na nightglow is its seasonal variation. The radiance shows a pronounced winter (both hemispheres) maximum. In the local summer it is difficult to observe because of the low level of its quantal output.

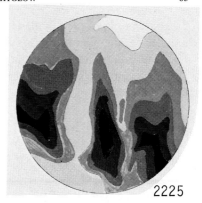

2225

4.3.2. THE NIGHTGLOW AND THE IONOSPHERE

Table 4-II brings together the facts concerning the nightglow radiation resulting from reactions involving ionized components of the upper atmosphere. The ionosphere is the consequence of the family of reactions indicated schematically in Figure 4-2b. For the details of the complex physical processes occurring in the ionosphere and the influence of the ionosphere on radio propagation, the reader is referred to standard works such as Rishbeth and Garriott (1969).

For the purposes of this book it is sufficient to point out two facts – first, the concentration of ionized components between 100 and 500 km is low relative to that of the permanent atmospheric components (Figure 4-9); and second, the reaction coefficients may be very high, especially when both reactants are electrically charged.

An astronaut in a polar-crossing orbit would be impressed with the dramatic drop in visual brightness of the upper atmosphere as he crossed the auroral zones and

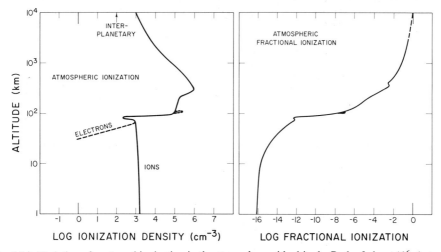

Fig. 4-9. Variation of nocturnal ionization in the atmosphere with altitude. Peak of about 10^6 electrons cm^{-3} represents a fractional ionization of only about 10^{-4}.

proceeded toward the middle latitudes and then on toward equatorial regions. The impression of a concentration of bright and dramatic optical displays in the auroral regions, and of contrasting faint and quiescent events in the lower latitudes is partly an illusion resulting from the fact that his eyes are not sensitive to the wavelengths where the low-latitude emissions are active. Especially interesting is the tropical activity of the O I red lines 6300/6364 Å, which we shall refer to generically as 6300 Å. At this wavelength the eye is only about 1/10 as sensitive as it is in the green.

The pioneering work in this field was done by D. Barbier (Figure 4-10), who in 1957 complemented his photometric studies at the Tamanrasset observing station in the

Fig. 4-10.   Daniel Barbier (1908–1965) with his multi-color nightglow photometer at the Haute Provence Observatory in Southern France.

Sahara with aircraft flights along the length of the African continent. He found from his ground observations that 6300 Å showed temporal changes. From his continental aircraft sweeps he found latitudinal changes. The ensemble of data gave a picture of two photometrically active regions about 20 deg on either side of the geomagnetic equator. The resulting photometric activity is a many-fold enhancement in brightness lasting over a period of an hour or two. The enhancement is superimposed on the 'typical' slow nocturnal decrease in the brightness of 6300 Å. After the well-defined photometric enhancement, the brightness returns to its earlier pattern of a slow decline in brightness (Figure 4-11).

That the source of the photometric perturbation in the brightness of 6300 Å is a local enhancement in the concentration of the ionized components is suggested by the

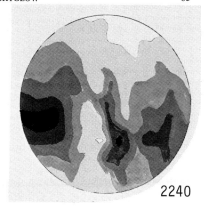

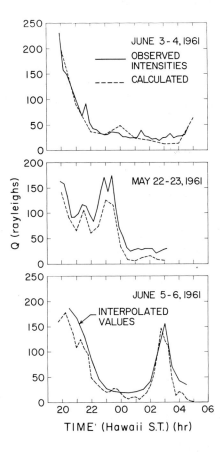

Fig. 4-11.   Three examples of nocturnal variations of 6300 Å at tropical station Haleakala (Hawaii). Upper plot is typical for nights when there is a steady decline in brightness during the night. Middle and lower plots illustrate cases in which significant interruptions in the photometric pattern occur. Dashed curves are computed by Barbier's Equation (4.8), using ionospheric parameters (see text) for comparison with observations (solid curves).

pair of reactions responsible for the photon emission:

$$O^+ + O_2 \rightarrow O_2^+ + O \tag{4.6}$$
$$O_2^+ + e \rightarrow O + O + 6.96 \text{ eV}. \tag{4.7}$$

The rate of the process is controlled by the first reaction. The photochemical reaction responsible for the emission (Equation (4.7)) is sufficiently exothermic to excite the O atoms to both the ^1D and the ^1S states, which action is consistent with the fact that both 6300 Å and 5577 Å are found in the tropical enhancements in the ratio $Q\ 6300/Q\ 5577 \approx 5$.

In order to deduce a relationship between ionospheric parameters and the quantal emission indicated by Equations (4.6) and (4.7), Barbier made use of the fact that ionospherists systematically characterize the state of the ionosphere by the use of two quantities that define the degree of ionization as well as the locus of its effective height: f_0F_2, the critical frequency of the F layer in the ionosphere, and $h'F$, the virtual height to the layer. The peak electron density n_0 (max) is related to the critical frequency by the relationship

$$n_0(\text{max}) = 1.24 \times 10^4 \ (f_0F_2)^2 \text{ electrons cm}^{-3},$$

where f_0F_2 is in MHz. The semi-empirical formula introduced by Barbier gave the quantal output of the emissions, Q, by

$$Q = A + B(f_0F_2)^2 \ e^{[-(h'F-200)/H]} \text{ R}, \tag{4.8}$$

where A and B are empirically determined constants and H is the atmospheric scale height in the emitting region.

Considerable effort has been expended to make more exact studies of the relationship between the ionosphere and the nightglow emissions in the tropics, but the Barbier formula has the advantage of giving a simple physical picture. Figure 4-11 shows three cases in which a comparison is shown between the Barbier formula and the observations of 6300 Å at the Hawaiian station at Haleakala (lat. N 20°43'; long. W 156°16'; geomagnetic lat. 21°N).

A typical tropical enhancement is relatively short lived during a given night, and at the end of the perturbation the photometric pattern continues as though there had been no interruption. A simple calculation using the prediction of Equation (4.7) indicates that the original local ionosphere should have been entirely 'burned out' by the loss of ionization during the enhancement. That the local ionosphere is *not* destroyed indicates that the phenomenon is probably associated with geostrophic movements of ionization, which bring 'clouds' of ionization to the particular tropical domain where the enhanced 6300 occurs. Here the reader might 'replay' the corner circle maps, which dramatically emphasize the enhancement.

An OGO (Orbiting Geophysical Observatory) satellite moving in an approximately polar orbit passes over the Earth's surface in about 90 min. Figure 4-12 shows a composite plot of the 6300 Å on a global basis, built up by a series of passes. This global representation is reminiscent of the pioneer work performed by Barbier in his flights

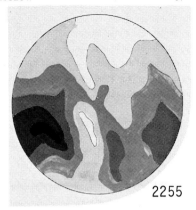

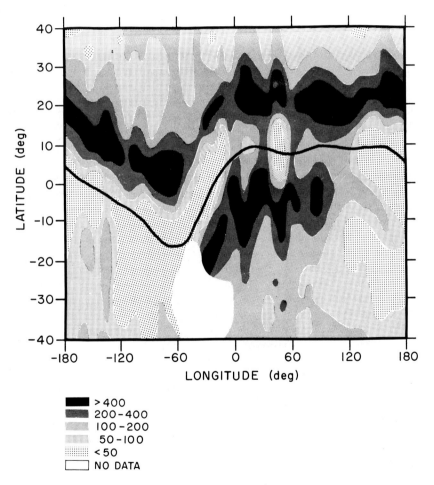

Fig. 4-12. Composite world-wide photometric map of 6300 Å made from a series of satellite passes (adapted from Reed *et al.*, 1973). The locus of the geomagnetic equator is indicated by a solid line, and the intensity scale, in R, is listed below.

along the African continent. It is apparent that the regions of concentrated 6300 Å emission are related to the geomagnetic equator. The photometric studies of the tropical nightglow have opened up a new domain of ionospheric exploration because they map large regions of the ionosphere in a synoptic way not possible from isolated ionosonde data.

4.3.2.1. *The Structure and Dynamics of* O I 5577 Å

In the previous section we discussed the temporal variations of 6300 Å nightglow in the tropics and mentioned that 5577 Å goes through a similar variation, not so large in amplitude and diluted photometrically. The major source of excitation of O I(^1S) results in an emitting layer of 5577 Å near the 100 km region, in contrast to the tropical ionospheric emission that occurs in the ionosphere above 200 km.

The 5577 Å emission, especially in midlatitudes where the ionospheric perturbations are not usually observed, goes through diurnal changes in brightness that have been much studied. Attempts to find large-scale representations of such changes in the form of circle maps have resulted in the following conclusions:

(1) The radiation occurs in discrete, cyclonic patterns.

(2) These patterns are in a sort of wind-like motion.

Maps (such as those in Figure 4-13) that have been arranged in cinema form (similar to the circle maps of 6300 Å) portray dynamic patterns that appear, traverse the circle of view, and then disappear beyond the observable horizon. Two or three such patterns may be identified during an observing night.

Whether these movements are actual winds or *in situ* changes that give the illusion of horizontal movement is not known. It is noteworthy that the wind concept is supported by the fact that the speeds of motion of the photometric features are similar quantitatively to the known wind speeds in the region of 5577 Å nightglow.

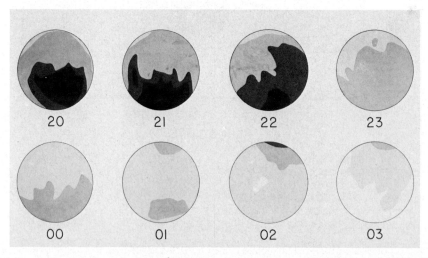

Fig. 4-13.   Series of circle maps of 5577 Å from observations made at midlatitude station of Fritz Peak, Colorado. Circle diameter is about 1000 km. Numbers indicate hours in Mountain Standard Time (7 hr W).

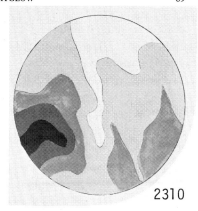

2310

4.3.3. The Nightglow Continuum

Whether there is a general spectoscopically structureless radiation from the upper atmosphere in the visual part of the spectrum is really a moot question. From the observational standpoint, the existence of such radiation is difficult to establish because spectroscopically it cannot be distinguished from the zodiacal light and the integrated starlight unless observing techniques could be devised of such high resolution that the individual Fraunhofer absorption lines of the astronomical components of the LONS could be resolved. The present evidence suggests that the so-called nightglow continuum is partly a real upper-atmosphere phenomenon and partly due to sunlight that has suffered multiple scattering in the atmosphere. Exact quantitative measurements from space will probably ultimately give the answer.

4.4. The Nightglow from Space: the Exosphere and Geocorona

We introduced this chapter by reference to the fact that the visual nightglow layer near 100 km, due chiefly to atomic O I 5577 Å, is a characteristic feature of the astronauts' look at the Earth's limb during satellite night. During the first full decade of the space age, the 1960s, there was, in parallel with the manned program, a strong program of observations from unmanned satellites and from rockets which yielded new information on the nightglow from topside.

Referring back to Figures 4-3 and 4-4, we note that the upper atmosphere from 1000 to 10000 km above the Earth thins out to a density approaching that of interplanetary space. In this region, called the exosphere, kinetic collisions become very infrequent. The chemical composition changes to a H–He mixture and finally to an ionized plasma (Figure 4-9). At 1000 km, the frequency of collision is only about one in 10 min (contrasted to about 4000 s^{-1} at 100 km). A particle travels on the average about 500 km between collisions, and the statistical probability of the escape of a particle from the gravitational hold of the planet without any collisions with other particles becomes significant.

In this region and continuing out into a still rarer domain, a sort of geocorona has been observed, consisting of emissions from the H atom (the Balmer α line, 6563 Å, and the Lα, 1216 Å, and β, 1026 Å, lines) and from the He atom emissions at extreme wavelengths of 584 Å in the UV and 10830 Å in the IR. Also, an ionized He emission at 304 Å has been isolated. The UV emissions can be observed only from rocket or satellite above the lower absorbing atmosphere. Other emissions that have been observed from space are the permitted O I lines, 1304 Å and 1356 Å, which are iono-spherically controlled and, like the 6300/6364 Å lines, are characteristically concentrated in the tropics.

4.5. Midlatitude Stable Auroral Arcs (M/SAR)

In 1958, one year after his report on the tropical enhancements of O I 6300 Å, Barbier, working at the midlatitude station of the Haute Provence Observatory, announced the discovery of an arc-like phenomenon that bears a superficial similarity to the tropical phenomenon discussed earlier (p. 64), in that the same radiation, 6300 Å, is involved. It differs in other respects, however, indicating that it must have a quite different physical origin. Barbier dubbed it *L'arc aurorale stable* and it has come to be known as a SAR arc, an anglicized acronym of Barbier's appellation. It is highly selective spectroscopically, favoring strongly 6300 Å, and it is a midlatitude phenom-enon – hence it is also sometimes referred to as an M/SAR arc (monochromatic; midlatitude).

Its energetic selectivity, ^1D (1.96 eV) over ^1S (4.17 eV) (see Figure 4-8) is in marked contrast to the general spectroscopic nature of the polar aurora, which is definitely a high-energy phenomenon in terms of the excitation levels of the emissions. The M/SAR-arc is energetically different from the tropical case, in which the green line is enhanced along with the red.

M/SAR arcs are observed during times of geomagnetic activity and are clearly as-sociated with auroral displays, though discretely to their equator side. They are ori-ented along geomagnetic parallels and hence must be associated with moving charged particles in the upper atmosphere. If the association is one of *direct* excitation of resi-dent atoms by incoming charged particles, it becomes necessary to impose a strong upper limit on their energy such that the resident oxygen atoms are excited to the ^1D state (1.96 eV) but not to the ^1S state (4.17 eV). What seems more likely is that the ex-citation is caused *indirectly* by the raising of the temperature of the resident atoms.

In Figure 4-14 we show schematically the zones in which the three phenomena occur – the polar aurora, A; the M/SAR arcs, M; and the equatorial perturbations, E. The M zone is shown as separated from the A zone because, although the aurora may on rare occasions extend to equatorial regions, the M arcs have usually been observed discretely and equator-ward with respect to coincident polar auroras (Figure 4-15). The M/SAR arcs have their photometric center near 400 km (Figure 4-16) in contrast to the polar aurora, which extends down to the 100 km region. The geomagnetic field is obviously in control, as evidenced by the orientation of the photometric maxima along the geomagnetic-latitude circles.

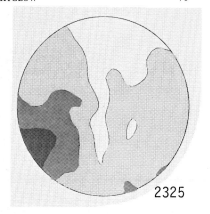

2325

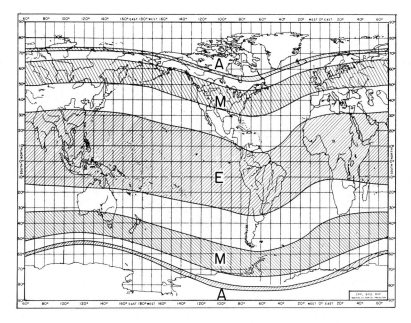

Fig. 4-14.   Schematic representation of terrestrial distribution of equatorial nightglow (E), midlatitude
arcs (M), and polar aurora (A).

From a single station, the M/SAR arcs can be observed over several thousand kilometers. On one occasion, an arc was observed in France during one night and a similar one along the same geomagnetic latitude in Colorado the same night immediately after the close of the observing in France, indicating that an excitation ring extended at least one third of the way around the globe. The absolute radiance of the arc has been measured as high as 8000 R. If this were at a wavelength near 5577 Å, the phenomenon would be a conversational spectacle, but the visual threshold of the eye in the red is about 20 000 R (only 1000 in the green), so this dramatic display of

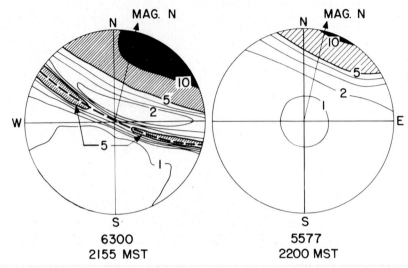

Fig. 4-15.   Photometric pattern of an M/SAR arc as observed at Fritz Peak, Colorado. In contrast to the well-defined arc for 6300 Å, 5577 Å does not show the phenomenon, although both wavelengths display the conventional aurora near the horizon, toward magnetic north.

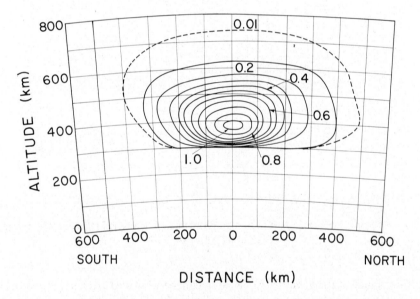

Fig. 4-16.   Composite cross-sectional isophotal map of 6300 Å M/SAR arc (according to Tohmatsu and Roach, 1962).

nature must be 'seen' by instruments. If there were a small upward shift of 0.2 eV of the ^1D level, then the $^3P \leftarrow {}^1D$ radiation would be green, and the M arcs would be a source of great beauty. Of course, such a shift would move 5577 Å ($^1D \leftarrow {}^1S$) into the red, and the appearance of the polar aurora would be drastically changed.

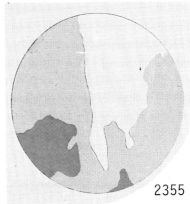

2355

Since much of our research motivation has been initiated by the excitement arising from visual observations, we may conclude that a mere 0.2 eV in a single atomic-energy level has affected nightglow/aurora scientific history!

4.6. The Polar Aurora

The polar aurora differs from the nightglow in two observational features – one in degree and the other in fundamentals. First, the aurora is brighter than the nightglow,

Fig. 4-17.   Photograph of polar aurora, looking toward the (geomagnetic) east. Reproduced from a color slide taken by Dr John Boyd of the Geophysical Institute, University of Alaska (with permission).

especially in the visible part of the spectrum (Figure 4-17). Second, the spectroscopic features of the aurora indicate a high level of excitation, being due to the transfer of energy from high-speed incoming particles, in contrast to nightglow excitation arising from the exchange of energy in photochemical reactions. The M/SAR arc is, in this respect, an intermediate phenomenon – it is a low-excitation case but is usually classified as an aurora because of its coincidence with auroras and its geomagnetic alignment. The comparative energetics of the upper-atmosphere emissions discussed in this chapter are shown in Table 4-III. The auroral brightnesses progress from International Brightness Coefficient (IBC) I, for which the vertical brightness of 5577 Å is 1000 R (1 kR) by decade steps to IBC IV, for which the 5577 Å brightness is 1000 kR. The summation in Table 4-III of *all* the polar aurora features is taken from Chamberlain (1961:197, table 5.5).

TABLE 4-III

Energetics of upper-atmosphere emissions

Phenomenon	Total photon emission (R)	Total energy emission[a] $(\text{ergs cm}^{-2}\,\text{s}^{-1})$
Nightglow		
OH	4 500 000	3.2
Other	88 000	0.125
M/SAR Arc	8 000	0.025
Polar Aurora		
IBC I	68 200	0.18
II	682 000	1.8
III	6 820 000	18.0
IV	68 200 000	180.0

[a] In a column.

The visual observer of the aurora often sees a dynamic picture of apparently rapid movements across the sky as the locus of the excitation changes. Colors are evident for auroras between IBC II and IBC III as the *color* threshold of the eye (about 20 000 R) is exceeded. Mixes of O I 5577 Å, O I 6300 Å, Hα 6563 Å, and N_2 first positive bands contribute to a variety of color sensations. If 5577 Å predominates, the color is green; O I 6300 Å, when present, often produces a red upper overlay. The first positive bands of N_2 occur below the aurora proper, and when bright enough produce a red lower border. Even a pastel shade may result from a mixture of several radiations.

Polar auroras tend to occur in the so-called auroral zones. In addition to this spatial preference, there are well-defined temporal fluctuations. There are maxima of occurrence in the spring and autumn (Figure 4-18). Auroral activity goes through an 11 yr cyclical variation associated with the activity on the Sun as measured by sunspot numbers (Figure 4-19). The auroral maximum in general occurs about 2 yr after the sunspot maximum.

0025

The excitation mechanism of a polar aurora is probably the direct excitation of the atmospheric constituents by charged particles proceeding along geomagnetic lines of force to the portion of the atmosphere where impacts are sufficiently numerous to decelerate the intruders, with multiple excitations resulting from the impacts. Bright auroras are of great beauty and are often dramatically variable in position, form, and color as the excitation conditions in their variability often give the impression that the auroral features are sweeping across the sky in a 'wind-like' motion of great speed.

When present, these brilliant displays are striking features of the sky. They preclude many astronomical studies for which a sky with a faint background (or foreground) is necessary, however. Therefore we take advantage of the fact that polar auroras tend to be concentrated in fairly restricted zones and occur somewhat sporadically

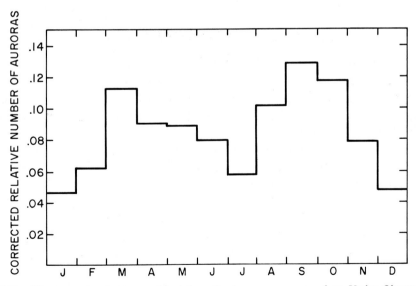

Fig. 4-18. Histogram showing seasonal variation of polar auroras as seen from Yerkes Observatory, corrected for cloudiness and number of dark hours each month (from Meinel *et al.*, 1954:411).

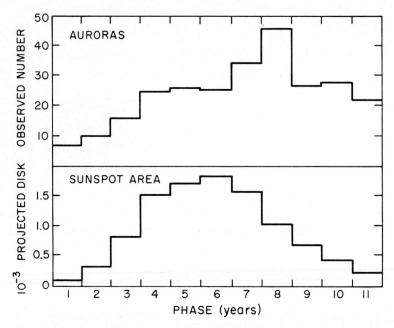

Fig. 4-19. Variation of auroral occurrence as seen from Yerkes Observatory during a sunspot cycle, based on a composite over several sunspot cycles (from Meinel *et al.*, 1954: 408). Zero phase refers to sunspot minimum.

in giving them only a comparatively brief treatment. The reader interested in an understanding of the complexity and the physical origin of the auroras should consult such books as that of Chamberlain (1961).

4.7. Effect of Nightglows on LONS Studies

The upper-atmosphere emissions discussed in this chapter present a general photometric foreground that limits the astronomer in his examination of the very distant objects which, because of their distance, are very faint. The spectroscopically discrete radiations can be avoided by the use of filters that exclude them. But there is a residual nightglow that is spectroscopically continuous over much of the spectrum. This cannot be eliminated by filtering and constitutes a general limitation on deep probing of the universe. In recent years it has become possible to overfly the nightglow via satellite, a procedure which not only eliminates the nightglow foreground light but also permits an extension in wavelength into both the UV and the IR spectral regions, which are seriously absorbed by the lower atmosphere.

Astronomers think of our planet as an observing platform and grudgingly accept the limitations imposed by its atmosphere, because of the mitigating circumstances of its stability and livability. Upper-atmosphere physicists, on the other hand, take a more parochial view of the situation, applauding the nightglow because of its contribution to an understanding of upper-atmosphere processes, accepting with some

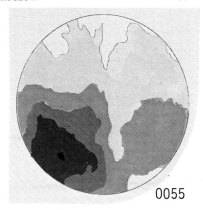

0055

grace the technological difficulties imposed by the presence of light of astronomical origin in the night sky.

Appendix 4-A. The Change in Brightness of a Nightglow Layer with Zenith Distance

We have mentioned that a distinguishing feature of a nightglow layer is its increase in brightness from the observer's zenith toward his horizon. This change is due to the fact that an emission occurring in a uniform layer bounded by concentric arcs with respect to the center of the Earth will present an emission path length to an observer on the Earth's surface increasing with zenith angle, z, according to *

$$V = \left(\sqrt{1 - \left(\frac{r}{r+h}\right)^2 \sin^2 z} \right)^{-1},$$

where r is the radius of the Earth and h is the height of the emitting layer above the Earth's surface. Some values of the function V are shown in Table 4-IV, with z and h as parameters.

TABLE 4-IV

The van Rhijn function, V, at different altitudes

Zenith angle (deg)	V at different altitudes			
	100 km	200	300	400
0	1.000	1.000	1.000	1.000
40	1.292	1.279	1.267	1.256
60	1.914	1.841	1.779	1.725
70	2.635	2.426	2.267	2.141
80	4.087	3.367	2.945	2.661
90	5.713	4.086	3.374	2.955

* The equation was first used by P. J. van Rhijn (1921) in an atmospheric connotation. The function, V, is often referred to as the van Rhijn function.

Early attempts to utilize the zenith-to-horizon increase of nightglow brightness as a height determinant were of minimal value because of: (1) difficulties in making proper allowance for the extinction and scattering of the lower atmosphere and (2) the fact that the nightglow as scanned by a given observer is intrinsically patchy, thus violating the basic assumption of a uniform emitting layer used in deriving the V equation.

For a discussion of these effects, see Roach *et al.* (1958). Actually, our present knowledge of nightglow-layer heights is based largely on their direct observation from rockets that traverse the layers.

References

Barbier, D.: 1957, *Compt. Rend.* **244**, 1945.
Barbier, D.: 1958, *Ann. Géophys.* **14**, 334.
Chamberlain, J. W.: 1961, *Physics of the Aurora and Airglow*, Academic Press, New York.
Chapman, S.: 1931, *Proc. Phys. Soc. (London)* **43**, 26; 483.
Elvey, C. T.: 1950, *Am. J. Phys.* **18**, 431.
Hennes, J. and Dunkelman, L.: 1966, *J. Geophys. Res.* **71**, 756–757.
Meinel, A. B., Negaard, B. J., and Chamberlain, J. W.: 1954, *J. Geophys. Res.* **59**, 407.
Rayleigh IV, Lord (R. J. Strutt): 1919, *Astrophys. J.* **50**, 227.
Rayleigh IV, Lord (R. J. Strutt): 1921, *The London Times*, 6c, 27 Jan.
Rayleigh IV, Lord (R. J. Strutt): 1930, *Proc. Roy. Soc. (London)* **A129**, 458.
Rayleigh IV, Lord (R. J. Strutt) and Jones, H. S.: 1935, *Proc. Roy. Soc. (London)* **A151**, 22.
Reed, E. I., Fowler, W. B., and Blamont, J. E.: 1973, *J. Geophys. Res.*, in press.
Rishbeth, H. and Garriott, O. K.: 1969, *Introduction to Ionospheric Physics*, Academic Press, New York.
Roach, F. E., Megill, L. R., Rees, M. H., and Marovich, E.: 1958, *J. Atmospheric Terrest. Phys.* **12**, 171.
Roach, F. E. and Smith, L. L.: 1967, in B. M. McCormac (ed.), *Aurora and Airglow*, Reinhold Publishing Corp., New York.
Tohmatsu, T. and Roach, F. E.: 1962, *J. Geophys. Res.* **67**, 1817.
van Rhijn, P. J.: 1921, *Publ. Astr. Lab. Groningen*, No. 31, pp. 1–83.
Yntema, L.: 1909, *Publ. Astr. Lab. Groningen*, No. 22, 1–55.

0125

PLATE V   ROBERT JOHN STRUTT, LORD RAYLEIGH IV

1875–1947

Sons of renowed fathers often have trouble establishing their own legitimate claim to fame, particularly when they are working in the same field, as did the Rayleighs. Fortunately, mutual respect and affection overrode rivalry in this family, and Robert John Strutt built capably on concepts developed by his father, Lord Rayleigh III, branching out into creative research of his own as he matured. His major contributions in the aspects of physics of the atmosphere led to his becoming known as 'the airglow Rayleigh'.

Born on August 25, 1875, Robert John Strutt was the eldest child of Lord Rayleigh III and his wife, the former Evelyn Balfour. Strutt's formal schooling included The Grange, which his Balfour uncles had attended; Eton, in which his interest in and experience with science expanded but where he diligently avoided making friends and participating in sports; and Trinity College, Cambridge, which he entered in 1894. There he started in Mathematics but soon found the Natural Sciences more to his liking. In this latter area at that time, emphasis was on experimental physics, which was taught at Cavendish Laboratory, then directed by J. J. Thomson, a former pupil and close friend of Rayleigh III. There Strutt learned laboratory skills – such as lathe turning and glass blowing, which were necessary because very little ready-made equipment was available – in addition to developing his considerable talents in designing and carrying out experimental problems and devising techniques and equipment to obtain data. During all his school years, he corresponded enthusiastically with his father about the latter's experiments, eagerly inquiring as to what had happened, and how, and why.

Perhaps it was his extensive correspondence – first with his father and later with other scientists and intellectual contemporaries – or perhaps an inborn talent, but in any case, Robert Strutt was a fluent writer and, throughout his life, held in high regard the ability to write concisely and lucidly. His scientific papers were models of excellent exposition, and his frequent forays into more widely read publications – such as letters to the editor of the *London Times* and statements to the House of Lords on matters having scientific aspects – were phrased for clear understanding by the layman who would be their audience. He also wrote two outstanding books – one a biography of his father, Lord Rayleigh III, and the other a biography of J. J. Thomson, his teacher.

Strutt was elected a Fellow of Trinity College in 1900, just a year after the first of his many scientific papers was published in the *Proceedings of the Royal Society*. In 1905, when only 29, he was elected a Fellow of the Royal Society. That was a momentous year for him, as he married Lady Hilda Clements, daughter of the 3rd Earl of Leitrim of Donegal, Ireland. They lived in Cambridge until 1908, when Strutt was appointed Professor of Physics at the Imperial College of Science and Technology in London.

Probably because of his father's work with photography – in the days of exacting requirements of wet collodion plates – Strutt became interested in photography and acquired skills that later were invaluable to many people. During World War I, he learned that the healing of shrapnel-wounded soldiers in a nearby

Lord Rayleigh IV

(Photograph courtesy of John H. Howard, Air Force Cambridge Research Laboratories, Laurence G. Hanscom Field, Bedford, Massachusetts.)

hospital was impeded because the doctors could not locate the shrapnel splinters to remove them. Volunteering his services, he took his equipment to the hospital, X-rayed the patients in their wards, developed the plates in a temporary darkroom nearby, and turned the plates over to the surgeons, who could then remove the remaining splinters. Recognizing a more general need for this service, he volunteered to carry out the same work with the British forces in France and, having moved his family to Terling, the family estate, he was sent to France in June 1915. He was soon recalled to London, however, to serve as a member of a panel working with the Inventions and Research Board to assess the equipment requirements of Admiralty Departments and to evaluate suggestions and inventions submitted to fill those needs. He continued in this capacity until the end of the war.

In April 1919, Lady Hilda, who had been in poor health for several years, died; this loss was a severe blow to Strutt and left him with the care of their four children – three sons and a daughter.

Only a few months after his wife's death, Strutt was again saddened, by the loss of his father, Lord Rayleigh III; at that time, the title, with its privileges and responsibilities, descended on his shoulders. Officially now the master of Terling, he instituted a modernization program, installing an electrical system through the buildings – only his father's laboratory had been served by electricity prior to that time.

Late in 1919 Strutt, now Lord Rayleigh IV, met a war widow, Mrs Kathleen Cuthbert, through mutual friends; they were married in July 1920. Mrs Cuthbert had four children also, of about the same ages as Rayleigh's, and they later had one son, giving them a total of nine children. Rayleigh acquired a car with a special body that accommodated all eleven members of the family, for short trips around the countryside. For more extensive journeys, the family traveled by train and hired bus.

For some time Rayleigh had been interested in the light of the night sky – his father's early paper on scattering of light of the daytime sky may have given him the initial impetus. In 1919 he published his first

0155

paper on the subject, titled 'Polarization of the Night Sky' (Rayleigh IV, 1919), which was followed over the next 16 years by numerous additional papers on more intensive research, covering observations and measurements of the night-sky spectrum. His early-1920s observations by photographic and visual photometry of the light of the night sky gave quantitative data on time variations of intensity in different spectral regions. On the basis of the 5577 Å green line present even during periods of no auroral activity, Rayleigh was convinced that there was a terrestrial component of the night-sky light that was not auroral in origin, and he designated this component as the nonpolar aurora. In 1930, he presented his paper expressing the radiance of the green (airglow) line in absolute quantal units (Rayleigh IV, 1930), which subsequently resulted in the adoption of the *rayleigh* in his honor (see Appendix 2-B). In the early 1930s he worked with H. Spencer Jones, then Astronomer Royal, further refining his analysis; their joint paper (Rayleigh IV and Spencer Jones, 1935) was his final published contribution in this field.

Rayleigh was widely recognized as a skilled experimenter, ingenious at designing simple yet definitive experiments and devising techniques and equipment to test his theories. His conclusions were, therefore, firmly based and earned for him the respect and recognition of his scientific contemporaries over his half-century career.

Of his 321 published papers, more than half dealt with his extensive original research on gaseous phenomena; his valuable contributions to airglow knowledge comprise a relatively small proportion of that half. His remaining papers covered a wide range of subjects, to many of which he also made significant contributions. More than 50 years ago, Rayleigh published a paper of interest in the light of today's concerns on the part of environmentalists – on oil spills in the oceans and their potentially devastating effects (Rayleigh IV, 1921) – and in 1938 he wrote an article, unpublished, expressing his concern on overpopulation (Rayleigh IV, Ms.).

Thus Rayleigh IV was not a scientific recluse – he was an active participant in the contemporary English scene, enjoying family and social activities and accepting the responsibilities of public service, as well as exerting his scientific talents throughout his life. But he was a scientist to the end – he worked 10 hours in his laboratory on the day before his death, at age 72, in 1947.

CHAPTER 5

DUST-SCATTERED STARLIGHT – THE DIFFUSE
GALACTIC LIGHT

Integrated starlight, zodiacal light, and nightglow – which have been discussed to this point – are the major contributors to the terrestrial night sky light. The resolution of the total light into its principal components is much facilitated by the fact that each has a specific region of concentration in the sky, such concentration being apparent even to the unaided eye.

Component	Region of concentration
Integrated starlight	The galactic plane
Zodiacal light	The ecliptic, the Sun
Nightglow	The horizon

The spectroscopically discrete nightglow emissions can be differentiated from the background astronomical light by the introduction of spectroscopic resolution either in the form of photometers with narrow filters, whose peak transmission is centered on the emission feature, or by spectrometers, with resolutions in the 1 to 10 Å range. The nightglow continuum, however, is not separable from the astronomical components by spectroscopic methods; therefore, we must here return to the technique of the preferential region of concentration for its analysis.

The active workers in the photometry of the night sky have developed a feeling for the uncertainties in the published tables, such as Table 3-I in this volume. In particular, the percentage-error bars associated with the faint regions of the zodiacal light (near the ecliptic poles) are probably greater than in the brightest regions.

The purpose of bringing up these uncertainties in the accurate evaluation of the *major* components is to emphasize that serious difficulties are necessarily associated with the estimation of any minor contributor to the night-sky light, such as the so-called diffuse galactic light, a glow associated with the concentration of stars and the scattering of starlight by the dust in the Milky Way.

In an early overall study of the night sky, Elvey and Roach (1937) reported an excess of light in the Milky Way after allowing for that due to the summation of the contribution of the unresolved stars in the photometric observations. A theoretical study by Wang (1936) was published while the observational paper was in press, predicting that interstellar dust co-existing with the stars should scatter the starlight much as the interplanetary dust scatters sunlight to produce the zodiacal light. The obvious independence as well as the quantitative agreement of the prediction and the observations led to general acceptance of the existence of the diffuse galactic light.

In retrospect, it is evident that the numerical agreement of the two studies may have been somewhat fortuitous, since the uncertainties in the observational analysis were significant. However, the existence of the excess galactic light has been subsequently confirmed by several investigators.

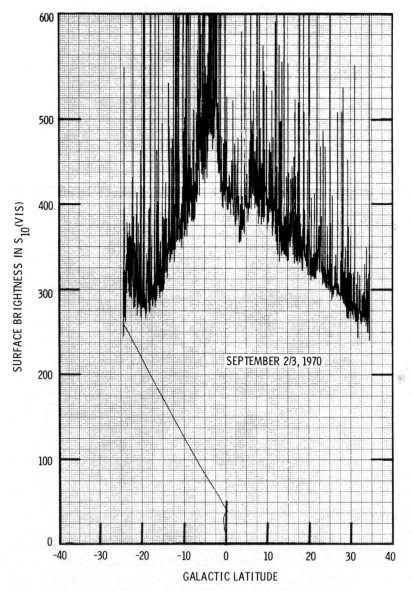

Fig. 5-1.   Plot of 2800 successive readings of sky brightness across the Milky Way with a fixed telescope. The Earth's rotation sweeps the astronomical sky across the field of view and, as individual stars enter it, discrete spikes appear on the plot. The underdrawn envelope tracing the lower points of the star deflections represents essentially the upper level of the diffuse galactic light superimposed on a general background of zodiacal light and nightglow.

The diffuse galactic light is brightest, in absolute terms, in the Milky Way, where both the interstellar dust that scatters the light and the concentration of stars themselves are at a maximum. If a large field of view is used, the contribution to the photometric reading from the integrated starlight is very strong, as is evident from even a casual examination of the heavy concentration of stars (Figure 2-2b) in the star-rich regions of the Milky Way. Many of the historical papers dealing with the problem have attempted to allow for the integrated light from the stars by a direct or, more often, indirect application of star-count data such as discussed in Chapter 2. The uncertainty introduced into the analysis by an application of this differential-residual method is severe, and we shall concentrate in this chapter on some recent results in which the employment of astronomical telescopes has permitted the use of small fields for which it is possible to make readings between most of the stars and thus minimize the correction for the general starlight.

To illustrate the nature of the problem, we refer to Figures 5-1 and 5-2. Figure 5-1 is a plot of a series of some 2800 digitized readings made with a photometer attached to a 16 in. telescope at Kitt Peak, Arizona. The telescope was fixed in position at a

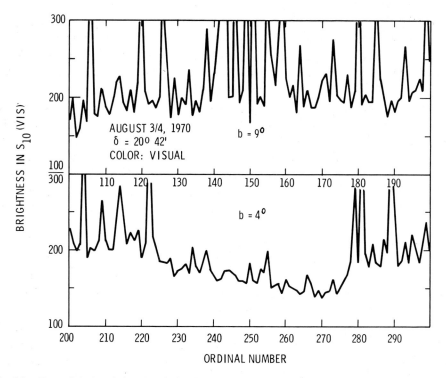

Fig. 5-2.   Expanded plot of observed sky brightness of a small portion of the domain shown in Figure 5-1 − between galactic latitudes (*b*) 9° and 4° − using a fixed telescope. The readings were taken at 15 s intervals through an entire night; 200 contiguous readings are shown in this plot. Star deflections again appear as discrete spikes. In this case, the zodiacal light and nightglow have been subtracted out and the curve represents primarily the diffuse galactic light, with a small but not trivial contribution from interstellar light from stars too faint to record individually.

declination of $20°42'$ and an hour angle of 2 hr W. The observations have been ordered in galactic latitude, and the brightness as the Milky Way traverses the field is apparent. The evaluation of how much of the maximum is due to diffuse galactic light has been simplified by the use of a relatively small field*. Figure 5-2 gives the detailed plot of 200 contiguous readings of a small part of the region covered in Figure 5-1, showing the detailed photometric structure of a portion of the Milky Way. In Figure 5-2 the zodiacal light and the airglow have been subtracted out, in contrast to Figure 5-1, for which much of the general background light is due to these contributors.

In a traverse of the Milky Way the concentration of stars maximizes, as does the background light, as shown in Figure 5-1, in which the numerous star deflections stand out above the underdrawn envelope. The star-deflection density was used to evaluate the correction necessary to allow for the residual light from very faint stars. A striking feature of the analysis is that the region of the slight dip in the readings near galactic latitude $+4°$ is actually brighter than its environs because the star correction here is almost trivial. The method outlined is now being used with fields so small that the correction for residual integrated starlight is trivial.

Another very satisfactory, but laborious, approach to the problem was introduced by Witt (1968), who carefully selected regions of the sky devoid of stars brighter than visual magnitude 20 for photoelectric observations across two regions of the Milky Way – one in the constellation Cygnus (galactic longitude $65°$ to $90°$) and the other in the Taurus-Auriga region (galactic longitude $165°$ to $190°$).**

Figure 5-3 shows plots of the regions observed by Witt together with the traverses of two sweeps across the Milky Way in the Cygnus region by the fixed-telescope method with automatic digitization of the readings. Inspection of Figure 5-3 indicates that the curves of $\delta = 31°58'$ and $20°42'$ cover rather closely the region covered by Witt's Cygnus observations. A comparison of the two independent results (Figure 5-4) shows that they are in substantial agreement for the Cygnus crossing, where the peak of the diffuse galactic light is more than 100 S_{10} (vis). The good agreement of the two sets of data gives confidence in the validity of both results.

The good general agreement between these observations also suggests that the diffuse galactic light in this part of the Milky Way is of a more-or-less sheet-like character with, however, sporadic localized photometric perturbations such as evidenced by the broad, low pips in the stretch of near-absence of star deflections in the lower part of Figure 5-2 (between readings 230 and 270).

We have here given only a discussion of the empirical facts of the starlight scattered by diffused interstellar dust and, furthermore, the sample is an extremely small fraction of the total. It is impossible to generalize on the basis of this amount of evidence, but in the next chapter we propose to discuss some of the properties of the star-

---

* The reader should refer to the original paper for a description of the technique used (Roach *et al.*, 1972).
** Note that Figure 2-6 indicates that, in the case of Cygnus, we are looking along a spiral arm and that, in the case of Taurus-Auriga, across two arms.

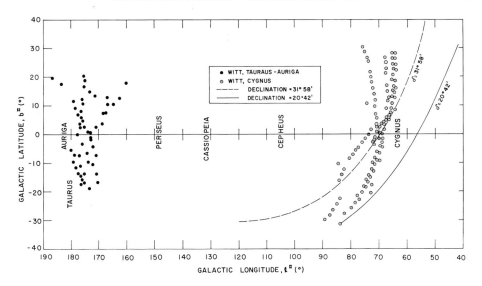

Fig. 5-3.   Plot of portions of the sky that have been studied for the presence of diffuse galactic light. The dot and dot-circle symbols represent regions examined by Witt (1968). Two regions covered by the fixed-telescope technique (Roach *et al.*, 1972) are indicated by the continuous plots.

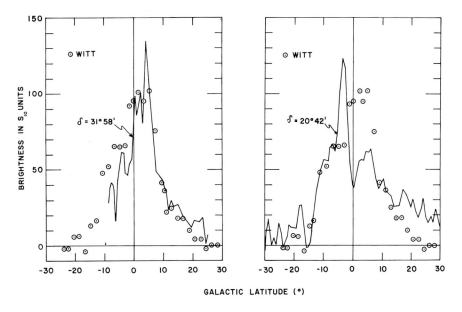

Fig. 5-4.   Comparison of independent measurements of diffuse galactic light by Witt (dot-circles) in the Cygnus region of the sky, and by Roach *et al.* (1972) (solid lines) by the fixed-telescope technique. In the latter, the effect of direct starlight is allowed for in the reduction of the data.

illuminated interstellar dust – including the functions of both the sheet and cloud nature of the interstellar medium in producing the diffuse galactic light – as well as the Sun-illuminated interplanetary (zodiacal) dust cloud.

References

Elvey, C. T. and Roach, F. E.: 1937, *Astrophys. J.* **85**, 213.
Roach, F. E., Smith, L. L., Pfleiderer, J., Batishko, C., and Batishko, K.: 1972, *Astrophys. J.* **173**, 343.
Wang, S. K.: 1936, *Publ. de l'Observatoire de Lyons, fasc. 19* **1**.
Witt, Adolf N.: 1968, *Astrophys. J.* **152**, 59.

CHAPTER 6

DUST – INTERPLANETARY AND INTERSTELLAR

In the physical universe, perturbative clumpings of material include the entire spectrum of sizes from the subatomic to the metagalaxy. The span of masses from the electron to our Galaxy goes from 10^{-27} to 10^{44} gm, or 71 powers of 10. Between these two extremes, Nature displays a comprehensive array to the astronomer, as listed in Table 6-I.

TABLE 6-I

Array of particulate matter

Species	Mass range (gm) (logarithmic)
Electrons	-27
Atoms	-24
Molecules	-23
Dust grains	-12 to 0
Pebbles, rocks, boulders (meteors, meteorites)	0 to 15
Comets	16 to 19
Asteroids	19 to 22
Moons	22 to 26
Planets	26 to 31
Stars	31 to 35
Galaxies (10^{11} stars like the Sun)	44

The fact that we are interested in dust in this book is not because it is, in itself, more significant astronomically than the other species, but because of its observability as a contributor to the light of the night sky. The observability of dust is somewhat fortuitous, depending first on its astronomical concentration, and second on the fact that the large number of small dust particles presents a larger cross section to light than do the smaller concentrations of large particles between dust and stars. Thus meteors, comets, asteroids, moons, and planets can be directly observed in the restricted region of the solar system, but their existence in interstellar space or associated with other stars must be inferred by indirect methods. The presence of interplanetary dust is indicated by the zodiacal light (Chapter 3) and of interstellar dust by the diffuse galactic light (Chapter 5). We make a digression to illustrate the statements of this paragraph.

According to studies of the dynamics of our Galaxy (Schmidt, 1965), the rotating system requires a mass distribution from the galactic center out to and beyond the Sun, according to Figure 6-1. Of particular interest is the requirement that at the

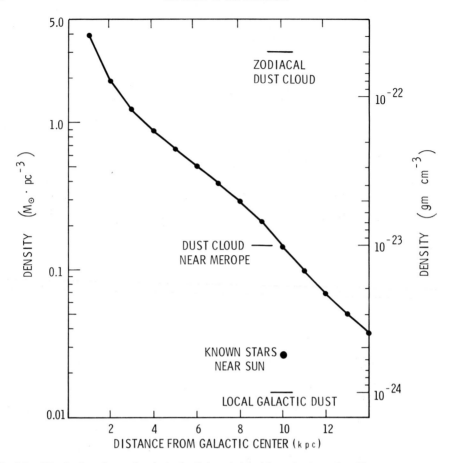

Fig. 6-1.   Distribution of mass density in the Galaxy deduced from the dynamics of its rotation according to Schmidt (1965). At the distance of the Sun from the center of the Galaxy (10 kpc) the density is 0.14 $M_\odot$ pc^{-3}.

Sun's distance from the galactic center (10 kpc) the density of material be 10^{-23} gm cm^{-3} or 0.14 $M_\odot$ pc^{-3}*.

Now, counts of stars (see Chapter 2) referred to the volume density in the astronomical vicinity of the Sun account for 0.02 $M_\odot$ pc^{-3}, or only 14% of that required by the galactic dynamicists. So we have a detective problem of astronomical proportions to locate the missing mass. It may be speculated that the several components listed in Table 6-I are distributed throughout interstellar space. A model based on such a speculation is presented in Figures 6-2 and 6-3. Figure 6-2 is basically an *interpolation* between the known space density of observable stars on the high-mass side of the plot to a model of the galactic-dust space density on the low-mass side. The implication of the plot is that *all* of the objects of Table 6-I occur throughout our part of the Galaxy. In Figure 6-3 we show the mass-density contributions (on a linear scale) according to

* $M_\odot$ = solar mass.

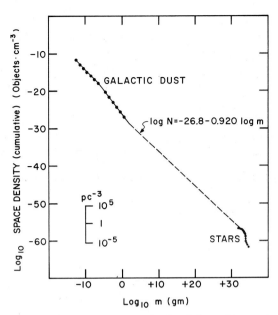

Fig. 6-2. An interpolation between the known space density of observable stars in the vicinity of the Sun and the interstellar dust cloud. The interpolated region includes all objects between these two extremes.

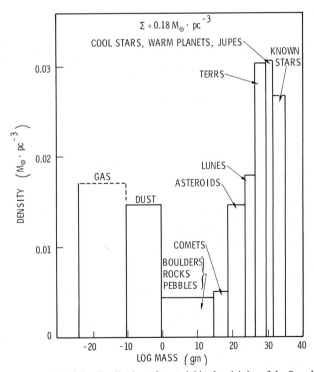

Fig. 6-3. A speculative model of the distribution of material in the vicinity of the Sun, based on the interpolation of Figure 6-2.

the masses of the several components. The total mass density of this speculative model is 0.18 $M_\odot$ pc^{-3}, in approximate agreement with the dynamic requirement*.

The model of solar-environment material just presented cannot be tested observationally insofar as the material with masses between dust and stars is concerned. Stars are directly and individually observable, being self-luminous. Interstellar dust is observable because of the cumulative effects of absorption and scattering over great astronomical distances. Figure 6-4 illustrates why the postulated astronomical debris with masses greater than dust cannot be detected. On a log scale is plotted the effective extinction coefficient, τ, for an assumed distance of 1 kpc, where τ is defined

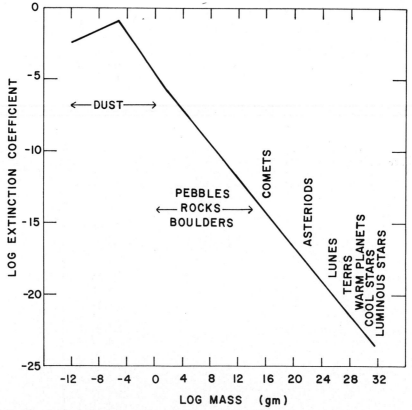

Fig. 6-4.   Plot of the geometrical cross section for an assumed distance of 1 kpc of the objects represented by the interpolations of Figures 6-2 and 6-3. In the plot it is assumed – for the purpose of illustrating the superiority of small dust particles in obstructing light traversing the medium – that the extinction coefficient τ is proportional to the geometrical cross section in the equation $I/I_0 = e^{-\tau}$. Actually, small dust particles are observable from interrelated mechanisms of extinction and scattering of light traversing the medium, the extinction causing a weakening of light from discrete objects such as stars, the scattering producing the phenomenon of diffuse galactic light.

---

* The reader interested in pursuing further the implications of the existence of interstellar debris might refer to Oort (1950), who considered the matter of a swarm of comets in the solar domain (halfway to Alpha Centauri), and Shapley (1962), who speculated on the general existence of cool stars and crusted planets in interstellar space.

by the following equation:

$$I/I_0 = e^{-\tau} \approx e^{-\pi s^2\, nl} = e^{-\sigma nl}. \tag{6.1}$$

The symbols are: I = the intensity of the transmitted light through the region, I_0 = the intensity of the original beam, s = the radius of the particle (cm), n = the space density of the particles (cm^{-3}), and l = the path distance (1 kpc = 3.09×10^{21} cm). We shall return to this later but wish here to emphasize (1) that as τ approaches unity $(\log \tau \rightarrow 0)$, as it does for the dust side of the plot, the extinction is significant*, and (2) that, where $\tau \ll 1$, as is the case for the larger-than-dust components, this family of objects cannot be observed by their cumulative extinction effects. Since the latter are not self-luminous, we have an observation gap between dust and stars.

Some amusing calculations of these elusive unobservables may be made. In the solar domain (defined as a sphere with radius halfway to Alpha Centauri), there are predicted six objects as massive as the planet Jupiter and 89 as the planet Neptune. Perhaps we should revise our concept of the morphology of our solar system. Within a sphere *including* Alpha Centauri, the model predicts five stars similar to the famous 'Barnard's star'. Searches for such intrinsically faint stars are, of course, part of the continuing efforts of astronomers.

The purpose of this general discussion of the total budget of astronomical material in our part of the Galaxy is to put into perspective the balance of the chapter, in which we shall discuss the zodiacal cloud of interplanetary dust and then the interstellar dust clouds associated with the diffuse galactic light.

6.1. The Zodiacal Dust Cloud

Referring back to Figure 6-4, we there used the geometrical cross section of the particles as an indication of their power to intercept light. The quantity τ, as so determined and used in Equation (6.1), is certainly valid for the large objects in estimating the extinction or absorption of a light beam traversing the region containing the particles. But the smaller particles cannot be considered as simple opaque objects with respect to the traversing light beam. The critical parameters are: (1) the size of the particle relative to the wavelength of the light, and (2) the index of refraction of the particles with respect to the light. A two-dimensional schematic array of the complex types of analysis is shown in Figure 6-5, taken from van de Hulst (1957). We have already encountered in Chapter 1 the case of Rayleigh scattering applicable in cases where the scattering particles of the atmosphere (molecules) are very small compared with the wavelength of the light being scattered. In general, the problems associated with interactions between the zodiacal dust cloud and sunlight and between the interstellar dust cloud and starlight are concerned with particles *not* small with respect to the wavelength of the light, a domain usually referred to as 'Mie scattering' after the investigator who concentrated on the problem.

---

* For example, $I/I_0 = 0.725$ for $\tau = 0.32$.

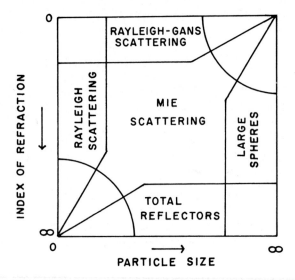

Fig. 6-5. Schematic diagram illustrating the complex of interactions between light and particles. The size of particle increases along the abscissa from very small to very large. The ordinate represents the index of refraction of individual particles, increasing downward (Adapted from Figure 20, van de Hulst, 1957: 132).

An understanding of the nature of the zodiacal cloud is approached not only by an interpretation of the photometric data on the zodiacal light but through such apparently diverse inquiries as: meteor studies (both optical and radar), meteorite investigations including their impact craters, and dust collecting and counting from space vehicles. All of these disciplines lead toward a knowledge of what has been called the 'meteoritic complex', small particles in interplanetary space, ranging in size from dust grains to small asteroids. According to Millman (1967), the basic character of the complex is given by a knowledge of its space density, velocity, and size distribution. Each of these properties is a function of distance from the Sun, time, and clumping. Finally, it is desired to know the nature of the material, such as composition, density, structure, hardness, and strength. It is obvious that such a large number of parameters cannot be determined explicitly from a restricted observational domain, or even from the ensemble of all the observational approaches. A specialist in any one of the disciplines makes a partial contribution to the interpretation, but a generalist is needed for the complete diagnosis.

The observations of the zodiacal light (and the gegenschein) help in estimating the space density near the Earth and, as a function of distance from the Sun, the size distribution of the particles, and their possible clumping.

There is a wide diversity of 'models' of the zodiacal cloud, as evidenced by the entries in columns 3 and 4 of Table 6-II from Bandermann (1967), which represent the extremes that have been postulated to account for the observational facts of the zodiacal light. A generally accepted definitive model between these extremes is not available, and we have chosen to illustrate the probable *general* nature of the interplanetary dust by the use of the parameters shown in column 5. Figure 6-6 shows the

TABLE 6-II

Comparison of models of zodiacal cloud

Item	Parameter	Bandermann A	Bandermann B	Illustration[a] (this book)
1	Typical size or radius, s	'Large'; several $\times 10^{-3}$ cm	'Small'; $< 10^{-4}$ cm	10^{-2} cm
2	Particle space density at 1 AU in ecliptic plane	10^{-21} gm cm^{-3}	10^{-23} gm cm^{-3}	2×10^{-22} gm cm^{-3}
3	Variation of space density with distance from Sun, r	$N \propto r^{-1.5}$	$N \propto r^{+0.5}$	$N \propto r^{-0.5}$
4	Distribution index, p; $N \propto s^{-p}$	'Steep'; $p > 4$	'Flat'; $p < 3$	$p = 4$ for $s > 10^{-2}$ cm $p = 1.35$ for $s < 10^{-2}$ cm
5	Composition	Metallic	Stony (dielectric)	
6	Average particle density, ρ, gm cm^{-3}	'Compact'; > 3 for stones; 7.8 for metals	'Fluffy'; < 0.1	1.0 for $s < 10^{-2}$ cm 0.44 for $s > 10^{-2}$ cm
7	Albedo, a	> 0.1	< 0.05 ('sooty')	
8	Shape	Nearly spherical	Needles; platelets	
9	Equilibrium electric potential	Small, + (a few volts)	Large, + or −	
10	Typical eccentricity of dust orbits, e	$\leqslant$ asteroids	Large, $e > 0.5$	
11	Average inclination of dust orbits	A few degrees	$> 20°$	
12	Typical geocentric speed	$5\left(^{+10}_{-2}\right)$ km s^{-1}	$\gg 15$ km s^{-1} or $\ll 5$ km s^{-1}	

[a] The examples chosen for illustration here are from Whipple (1967: 420, Table 3) for items No. 1, 2, 4, and 6, and from Gillett (1966) for item No. 3.

percentage distribution by mass according to the model along with a plot on a logarithmic scale of the 'age' of the particle in each mass regime according to Whipple (1967). Figure 6-7 shows the space density of the particles vs their radius according to the same model. In Figure 6-8 the relative geometrical cross section of the ensemble of particles is shown plotted against the mass on a logarithmic scale.

Finally, in Figure 6-9 is displayed the particulate space density in the ecliptic plane vs distance from the Sun on the assumptions: (1) a space density of 2×10^{-22} gm cm^{-3} at the Earth's distance of 1 AU, and (2) variation according to $r^{-0.5}$, where r is the distance from the Sun.

Immediately obvious is the fact that the present cloud cannot be a residual of the primordial material of the solar system. The mean life of the present cloud – 80 000 yr (Whipple, 1967; also see Figure 6-6) – is a trivial fraction of the age of Earth, and if it represents a steady state there must have been many thousands of turnarounds of zodiacal material during the history of our planet. Another generalization that can be made is that it is *not* possible that the zodiacal cloud is simply a part of the interstellar dust cloud. The space density of the zodiacal cloud is about 20 times the *total* required in the solar environment for dynamical stability of the rotating Galaxy (see Figure 6-1).

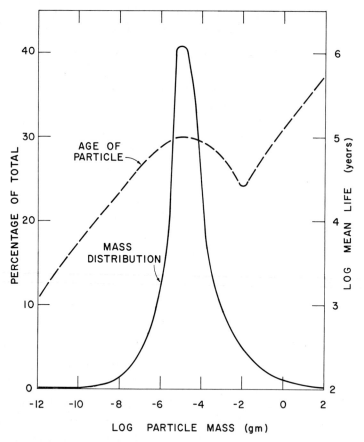

Fig. 6-6. The lifetime of the zodiacal cloud (according to Whipple, 1967).

If the present zodiacal cloud may be considered quasi-stable over a few centuries, it is pertinent to ask how it is maintained against the steady losses due to light pressure, spiraling-in (Poynting-Robertson effect), sublimation of particles, losses by solar wind, grinding by collisions, etc. We look for mechanisms that can supply some 10 to 20 metric tons of material per second. Two potential sources are the asteroids and the comets. The loss rate which has to be replenished corresponds to the grinding up of one asteroid such as Ceres (mass $\approx 7 \times 10^{23}$ gm) in about 10^{10} yr. Whipple has calculated that the mass loss of two periodic comets is comparable with the requirement for maintaining the zodiacal cloud – Halley's comet, 5 tons s^{-1}; Encke's comet, 3.5 tons s^{-1}. It is possible that *both* asteroidal and cometary contributions to the cloud are significant.

Of particular importance in evaluating the physical nature of the dust particles is the measurement of the polarization of the zodiacal light (see Figure 3-3). Early interpretations attributed this polarization to a concentration of electrons permeating the zodiacal cloud. That scattering from particulate material in the dust-grain category could produce polarization has led investigators into vigorous experimental and

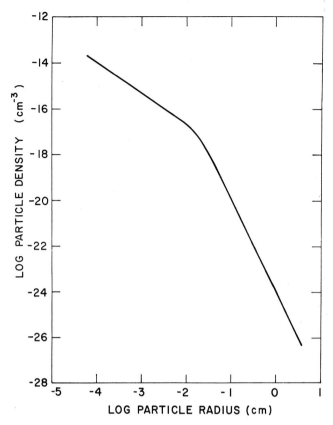

Fig. 6-7.   Space density of the zodiacal dust cloud with respect to the particle radius (according to Whipple, 1967).

computational attempts to rationalize the observations with predominantly dusty models.

Temporal and/or spatial clumping of the interplanetary material has been actively considered, especially since 1961, when Polish astronomer Kordylewski (1961) reported observations of photometric 'perturbations' near the so-called LaGrangian points of equilibrium in the Earth-Moon gravitational system. The mathematician LaGrange, in studying a restricted three-body problem, called attention to the fact that there are, in the plane of the orbit of two bodies, five regions of quasi-stability (Figure 6-10). Of particular interest are the points L4 (preceding) and L5 (following) at the apices of equilateral triangles formed by the two bodies. If there are no perturbing forces such as the presence of another massive body, then a nonmassive body in the L4 or L5 position will tend to be gravitationally caught in the equilibrium positions, perhaps performing oscillations (librations) around them. Applying this idea to the Sun-Jupiter system, astronomers have confirmed that there are more than a dozen asteroids (usually called Trojans and Greeks) at distances from the Sun equal to that of Jupiter and relatively close to the equilibrium positions.

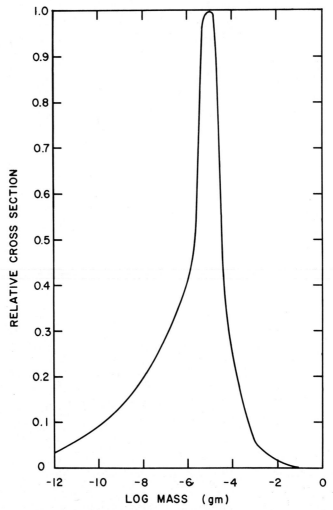

Fig. 6-8.   The zodiacal cloud according to Whipple (1967), showing the relative geometrical cross section vs the logarithm of the mass.

A careful search of one of the equilibrium positions in the Earth-Moon system for discrete asteroidal objects was unsuccessful, but interest in the possibility of a concentration of dust in these positions resulted from the report of Kordylewsky, and more recently by Simpson (1967). Recent studies suggest that the photometric perturbation resulting from concentration of dust in the Earth–Moon equilibrium positions may be as small as 10% to 20% of that of the gegenschein (Roach, 1973), which could account for the fact that some investigators have not been able to distinguish the effect (see, for example, Roosen, 1968).

The gegenschein has been ascribed to: (1) a phase effect – that is, a preferential *back*scatter of sunlight or zodiacal particles in the region beyond the Earth's orbit, (2) a concentration of particles at the L3 position of the Earth-Sun system, or (3) a

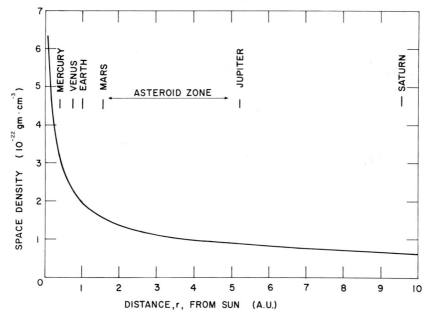

Fig. 6-9.   Plot of the space density of the zodiacal cloud on the assumption that it varies according to the
−0.5 power of its distance from the Sun.

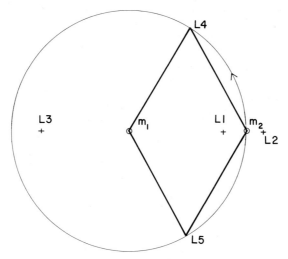

Fig. 6-10.   Schematic representation of the quasi-stable regions (LaGrangian points) for a small object
in the plane of two large rotating bodies.

combination of these effects. At the time of writing (September, 1973), preliminary
reports from the satellite Pioneer 10 indicate that the photometric gegenschein
persists well beyond the Sun-Earth L3 point. When the data are completely analyzed,
it should be possible to determine whether there is any degree of particulate concentration in the L3 region, as well as in the asteroidal belt. The latter is a possible primary

region in which large particles may be ground up into dust by collisions and thus contribute to the zodiacal cloud.

6.2. Interstellar Dust

As in the case of the zodiacal dust cloud, the evidence for interstellar dust is supported by several converging studies – the extinction of stars as a function of distance, the reddening of starlight that traverses either general or specific dusty regions, the polarization of starlight by intervening dust, and finally the diffuse galactic light described in Chapter 5.

The general gross visual appearance of the Milky Way as well as detailed studies

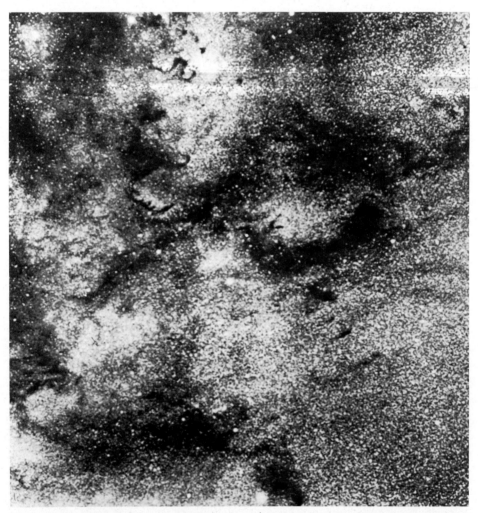

Fig. 6-11.   Photograph of a portion of the Milky Way near the star Theta Ophiuchi, showing many 'dark' regions due to intervening dust clouds that obscure the more distant stars.

show the effect of obscuration. The great rift in the Cygnus region is one of the features that catches the eye of the night watcher. Figure 6-11 shows a photograph of a region in the Milky Way in which the irregularities are obviously not due to perturbations in the distribution of stars but rather to the presence of intervening dust, which often has sharp delineations. Some of the very small blobs have been interpreted as regions where stars are in the birth process.

The general transmission of starlight looking outward from the solar system is $I/I_0 = Al = e^{-\tau}$, where I/I_0 is the fractional transmission of a star's light and l is the distance to the star in kiloparsecs, and A is in stellar magnitudes. In the vicinity of the Sun, $A \approx 0.35$ stellar magnitude/kpc. Thus in 1 kpc, starlight is weakened by 0.725. This very considerable diminution of a star's apparent brightness is accomplished by an extraordinarily thin density of dust extending over the vast distance of a kiloparsec $(3.09 \times 10^{21}$ cm$)$.

A somewhat oversimplified calculation will illustrate the space density of dust particles required to give an extinction of 0.35 mag. for a star at a distance of 1 kpc. It is convenient to take advantage of the sharp maximum in the relative cross section of particles in the Whipple zodiacal cloud at a mass of 10^{-5} gm (Figure 6-4) and apply it to the interstellar cloud. Order-of-magnitude calculations can be made using this as a 'standard particle'. With a particle radius of 0.0174 cm, the geometrical cross section of each particle is 9.5×10^{-4} cm^2. A 0.35-mag. weakening (transmission $= 0.725$) calls for

$$I/I_0 = 0.725 = e^{-\sigma n l},$$

where $\sigma =$ the cross section of each particle, $n =$ the space density of the particle, and l is the length of the path. For $I/I_0 = 0.725$, $\sigma n l = 0.322$, whence

$$n = 0.322/3.09 \times 10^{21} \times 9.5 \times 10^{-4}$$
$$= 1.1 \times 10^{-19} \text{ particle cm}^{-3} \ (1.1 \times 10^{-24} \text{ gm cm}^{-3} \text{ for } m = 10^{-5} \text{ gm}).$$

This corresponds to a space density of $(1.1 \times 10^{-24})/(2 \times 10^{-22}) = 0.005$ times that of the zodiacal cloud. The dust cloud around the star Merope in the Pleiades (Figure 6-12) has been estimated to have a mean density of 10^{-23} gm cm^{-3}, or 10 times that in the general interstellar medium (Figure 6-1). It is of interest to calculate the effects on light transmission for three cases, as shown in Table 6-III for several assumed distances. It is noted that a light beam 1 cm^2 in cross section is greeted by a standard particle in the general stellar domain every 10^{19} cm or 3.3×10^8 s (approximately 10 yr) (Figure 6-13). If such a cloud is in the immediate vicinity of our solar system, stars closer than 10 l-yr (c. 3 pc) could get a 1-cm^2 light signal to us without a single interruption by a standard dust particle until the zodiacal cloud was reached. Of such gossamer stuff is our Galaxy composed.

If the interstellar dust were as dense as the local zodiacal cloud (Figure 6-1), serious diminution of the brightness of even the nearby stars would occur. A star at 1 kpc distance would be completely invisible $(I/I_0 = e^{-60} = 10^{-26})$. A star at even the relatively close distance of 100 pc would be very seriously diminished in brightness $(I/I_0 = 0.0025)$.

Fig. 6-12.   Nebulosity (gas plus dust) around the stars of the Pleiades cluster. The star Merope is embedded
in the nebulosity in the lower part of the photograph.

On the other hand, the zodiacal cloud is of very low density in terms of solar-system distances. If its density in the vicinity of the Earth were to persist for 40 AU (to the orbit of Pluto) the weakening of the brightness of a star whose light traversed such a cloud would be only about 1% ($I/I_0 = 0.988$).

The general nature of interstellar-dust parameters based on the above considera-

TABLE 6-III

Extinction due to dust (standard particle, $\sigma = 10^{-3}$ cm^2) at different distances, l ($I/I_o = e^{-\sigma nl}$)

Cloud and its space density	1 AU ($l = 1.5 \times 10^{13}$ cm)			1 pc ($l = 3 \times 10^{18}$ cm)			1 kpc ($l = 3 \times 10^{21}$ cm)			10 kpc ($l = 3 \times 10^{22}$ cm)		
	nl	σnl	$e^{-\sigma nl}$	nl	σnl	$e^{-\sigma nl}$	nl	σnl	$e^{-\sigma nl}$	nl	σnl	$e^{-\sigma nl}$
Zodiacal cloud ($n = 2 \times 10^{-17}$ cm^{-3})	3(−4) 0.0003	3(−7)	0.999 999 7	60	0.06	0.942	60000	60	0	600000	600	0
Merope cloud ($n = 10^{-18}$ cm^{-3})	1.5(−5) 0.000015	1.5(−8)	1	3	0.003	0.997	3000	3	0.050	30000	30	9(−14)
Interstellar cloud ($n = 10^{-19}$ cm^{-3})	1.5(−6) 0.000 001 5	1.5(−9)	1	0.3	0.0003	0.9997	300	0.3	0.741	3000	3	0.050

[a] σ = extinction cross section of particle, cm^2.
n = space density of particles, cm^{-3}.
l = path length, cm.

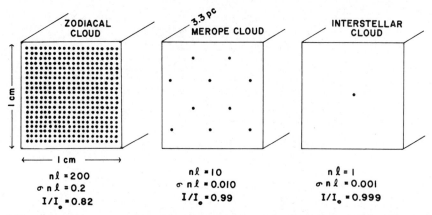

Fig. 6-13. Schematic representation of the relative number of dust particles included within a square-centimeter column extending over a distance of 3.3 pc (about 10 *l*-yr) for the zodiacal cloud, the Merope cloud, and the general interstellar dust cloud.

tion of the weakening of starlight leads to a discussion of its effect on the diffuse galactic light. Just as the interplanetary dust produces the zodiacal light by scattering sunlight, so dust-scattered starlight produces the diffuse galactic light described in Chapter 5. There is the difference that, in the first case, the observer has a definite position close to a single light source (the Sun) whereas in the second case he is surrounded by a large number of light sources (stars). In both cases the observer is embedded within the domain of the scattering material.

The diffuse-galactic-light observations have been interpreted in terms of two general parameters – the albedo of the particles and the angular dependence of the scattering function. For this latter, the scattering function, $S(\phi)$, is often referred to a parameter, g,

$$S(\phi) = \frac{1 - g^2}{(1 + g^2 - 2g \cos \phi)^{3/2}}. \tag{6.2}$$

In intuitive terms, g is the mean cosine of the scattering angle, ϕ. In Figure 6-14 we show graphical examples of g, including $g = 0$, corresponding to isotropic scattering, $g = 0.5$ and 0.6 strongly forward-scattering. For $g = +1$, the scattering is completely forward-scattering. Negative values of g refer to cases of predominant backscatter.

Two recent approaches to the analytical problem are based on the following assumptions.

(1) A dust sheet with a distribution similar to the stars in and near the galactic plane (van de Hulst and de Jong, 1969)

(2) Dust distributions in which discrete blobs occur, sometimes referred to as a raisin-pudding model (Mattila, 1970).

In both types of analysis, unique solutions are not possible, since a family of albedo, a, and g will usually satisfy the observations. However, when solutions for different regions are combined, the intersections have been interpreted as representative of the

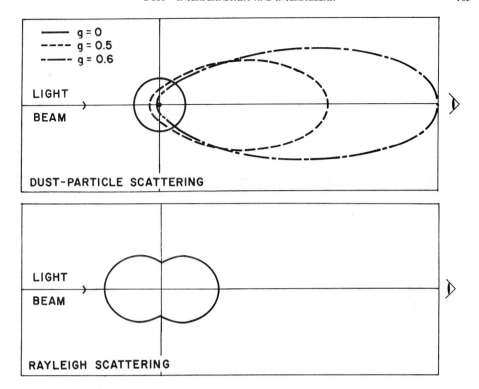

Fig. 6-14. Forward scattering of light by dust particles according to the parameter g (discussed in the text). In the lower part of the figure we repeat a part of Figure 1-2 to show the difference between scattering by dust particles and the very small (molecular) species responsible for Rayleigh scattering.

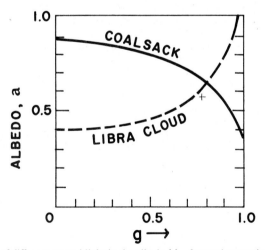

Fig. 6-15.   Solutions of diffuse scattered light in the albedo (a) – forward-scattering (g) domain according to Mattila (1970). The intersection of the solutions for the Coalsack and the Libra cloud at $a=0.65$ and $g=0.8$ is in close agreement with a similar analysis of observations of two regions in the Milky Way (Taurus and Cygnus) marked by the + ($a=0.55$, $g=0.75$), based on observations by Witt (1968); see also Figure 5-4.

physical nature of the interstellar medium*. Figure 6-15 illustrates the situation in in which several solutions have been superimposed on an a vs g plot. The general conclusion is that the albedo of the interstellar dust is near a midpoint value and that the material is highly forward-scattering ($g > 0.5$).

The early investigations of the diffuse galactic light led to an acceptance of interstellar dust as the scattering material causing the radiation, but the importance of this manifestation of matter in the astronomical scheme seemed as trivial as the size of the grains compared to such grand objects as stars. In recent decades the realization has grown that both ends of the mass spectrum (Figures 6-2 and 6-3) are not only important in our snapshot views of the universe but also that in an astronomical time scale there is an evolution of dust/gas $\rightarrow$ star $\rightarrow$ dust/gas. The *average* turnaround time of this process as dusty material coalesces into stars and stars live out their times and return to dust has been estimated to be 2×10^9 yr. Fortunately, our star-Sun has a longevity sufficient to permit us a spectator look at these cosmic events. During the 5×10^9 yr of our planet 'we' have witnessed some two-and-a-half stardust turnarounds. In a particular case it may not be apparent whether our snapshot picture of a dusty region is due to dust that has recently (astronomically speaking) been a part of an old star or is coalescing into a new one. In some cases it appears that discrete and well-defined blobs are 'birthing' stars.

References

Bandermann, L. W.: 1967, *Physical Properties and Dynamics of Interplanetary Dust*, Tech. Rept. No. 771. U. Maryland, Dept. of Physics and Astronomy, College Park, Md.

Gillett, F. C.: 1966, *Zodiacal Light and Interplanetary Dust*, NASA Contr. NsG-281-62 and NONR 710(22), Rept. 2.

Kordylewski, K.: 1961, *Acta Astron.* **11**, 165.

Kordylewski, K.: 1966, *International Astron. Union Circ.*, No. 1985.

Mattila, K.: 1970, *Astron. Astrophys.* **9**, 53.

Millman, P. M.: 1967, 'Observational Evidence of the Meteoritic Complex', in *The Zodiacal Light and the Interplanetary Medium* (ed. by J. L. Weinberg), NASA SP-150, 399.

Oort, J. H.: 1950, *Bull. Astron. Inst. Neth.* **11**, 91.

Roach, J. R.: 1973, private communication.

Roosen, R. G.: 1968, *Icarus* **9**, 429.

Schmidt, M.: 1965, 'Rotation Parameters and Distribution of Mass in the Galaxy', Ch. 22 in *Galactic Structure*, U. Chicago Press, Chicago.

Shapley, H.: 1962, 'Crusted Stars and Self-Heating Planets'. in *Revista*, Ser. A, *Matematica y Fisica Teorica*, Vol. XIV, No. 1 and 2, Facultad de Ciencias Exactas & Technologia Tucuman, Republica Argentina, 69.

Simpson, J. W.: 1967, *Physics Today* **20**, 39.

van de Hulst, H. C.: 1957, *Light Scattering by Small Particles*, Wiley, New York.

van de Hulst, H. C. and de Jong, T.: 1969, *Physica* **41**, 151.

Whipple, F. L.: 1967, 'On Maintaining the Meteoritic Complex', in *The Zodiacal Light and the Interplanetary Medium* (ed. by J. L. Weinberg), NASA SP-150, 409.

Witt, Adolf N.: 1968, *Astrophys. J.* **152**, 59.

---

* Navigators may see a similarity with the method of determining a terrestrial position by the intersection of plots of two line-of-position determinations.

CHAPTER 7

COSMIC LIGHT AND COSMOLOGY

Every thinking person is a cosmologist. The child ponders space and time and asks profound questions about their beginning and end. The adult scans the night sky and propounds the same questions, answers to which often include philosophical excursions. The optical astronomer makes quantitative measurements such as the red shift of the spectral lines of distant, extragalactic objects and notes that the shift increases with increasing distance of the object. From this simple (in principle) but difficult (in practice) measurement, a whole generation has become familiar with the concept of an expanding universe.

The study of the light of the night sky is an exercise in exact photometry, but beyond that it is an adventure in space and time that challenges the most nimble imagination. By far the largest spatial and temporal leap is the one under scrutiny in this chapter, involving, as it does, an attempt to evaluate the light from the totality of objects external to our Galaxy, the cosmic light. The establishment of the fact that many of the 'nebular' appearing telescopic objects are actually external to our Galaxy (island universes) is one of the outstanding achievements of the 20th century.

The measurement of the cosmic light from our terrestrial platform is complicated – almost obviated – by the photometric dominance of what, in this chapter, is the 'foreground' light – the nightglow, the zodiacal light, and the integrated starlight and the associated diffuse galactic light from our own Galaxy*. Here it is helpful to summarize typical values for the several competing components of the light of the night sky in the direction toward the galactic pole – the most favorable for the discrimination of the cosmic light (see Figure 7-1).

Understandably, attempts at measuring the cosmic light from Earth have not been

LONS component	S_{10}(vis) units
Nightglow	50
Zodiacal light	117
Integrated starlight	
from $m=5$ to ∞	28
$m=15$ to ∞	2.6
$m=20$ to ∞	0.15
Diffuse galactic light	9 (estimate)
Cosmic light	0.9 (de Vaucouleurs, 1949)

* From Earth's surface, the problem is further complicated by the difficulty of correcting for the extinction and scattering of our atmosphere, though this does not contribute a competing primary component.

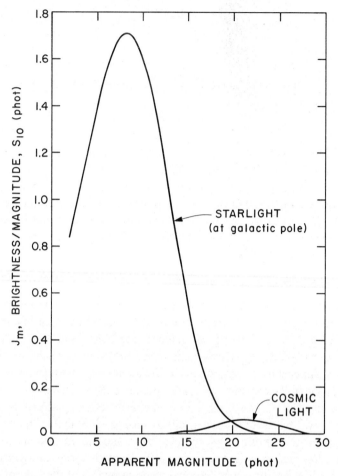

Fig. 7-1. Comparison of the light contribution from integrated starlight and from cosmic light as a function of apparent magnitude, in the direction of the galactic pole (adapted from de Vaucouleurs, 1949).

very successful. An empirical approach to the problem is to make counts of photographable nebulae to the observable limit and to extrapolate for the fainter objects. Such an estimate has been made by de Vaucouleurs (1949), who found that the surface brightness should be about 0.5 S_{10}(phot), corresponding approximately to 0.9 S_{10}(vis). Measurement of such a low level of brightness seems a difficult, if not impossible task. We recall the difficulty of measuring accurately the diffuse galactic light (Chapter 5) – the predicted brightness of the cosmic light is about 2% that of the diffuse galactic light. One attempt by Roach and Smith (1968) gave negative results – that is, a definitive value was not determined – but an upper limit of 5 S_{10}(vis) was reported (compare with the prediction of 0.9 by de Vaucouleurs).

There is possibly a more practical ploy that is being explored. The progressive red shift of the more and more distant extragalactic objects implies that the brightness of cosmic light relative to the foreground light increases as the observations are made at relatively long wavelengths. Thus, measurements in the IR could be fruitful.

It was the fond hope of many astronomers, during the planning decade before the start of the space age in 1958, that man's escape from Earth's surface would make possible an exact measurement of the cosmic light. In this chapter we shall explore this hope by way of a series of plots that will serve the double purpose of examining that possibility and at the same time taking us to the very horizon of the universe as presently known.

Any attempt to comprehend an overview of the great span of physical or temporal dimensions in our universe encounters the limitation of the human imagination, which is comfortable only with concepts in the general domain of human experience. The imagination deals satisfactorily with linear concepts – for example, plots in which both the x and y coordinates are directly proportional to the quantities under consideration. It is often impossible, however, to illustrate a particular domain by such a plot, and it is then frequently useful to make use of a logarithmic scale, in which successive integral steps on the plot correspond to a change by a factor, commonly ten.

Such a log plot is shown in Figure 7-2, in which we desire to take advantage of our recently acquired appreciation of the distance to the Moon by a plot showing how an astronaut en route to the Moon passes through the nightglow layer. On a linear plot drawn to scale, this would be unfeasible, since the Moon is some 60 R_E distant, and the nightglow is produced in an atmospheric 'skin' of a thickness of a few percent of an R_E. The nightglow-to-Moon span is conveniently shown in Figure 7-2, however, by the use of five logarithmic steps.

The unmanned satellites (such as OSO-6) have orbits at a height just above the nightglow at 400 km (Figure 7-2), but photometric observations from such a satellite might be affected by the nightglow, especially if the observation were made tangential to Earth. It is obvious, however, that current space technology makes possible observations from points sufficiently distant from Earth to be uncontaminated by the nightglow. Thus a first step has been taken in decontaminating our data toward the goal of evaluating the cosmic light. In one experiment alone (on OSO-6), several million observations of the 'night' sky have been made and are under intensive analysis.

We have placed two scales on Figure 7-2, giving the distance from Earth's surface in both kilometers and light-time (l-ms), as we wish to emphasize in this chapter the great expanse of our Universe not only in space but also in time. Note that the light-time from the portion of the upper atmosphere producing the nightglow is of the order of 1 ms and from the Moon about 1 s.

The excursions to the Moon in the manned space program, as well as the unmanned orbital satellites, have made possible the exclusion of the nightglow but have left in the data the zodiacal light, which, in large part, originates in a dust cloud extending from the Sun to envelop the Earth-Moon system and extend outward along the ecliptic plane. Figure 7-3 encompasses the part of the solar system in which the zodiacal cloud is concentrated. The zodiacal light, moving outward in the solar system, is assumed to

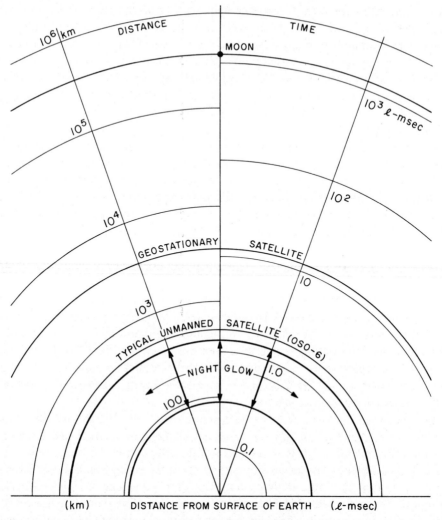

Fig. 7-2.   Schematic representation of near space on a logarithmic scale. Distances are indicated on the
left and light-time on the right.

decrease in brightness because of the probable decrease in the local density of dust as
we proceed out past Mars toward Jupiter (see Figure 6-9), and also to the fact that the
solar illumination on the dust decreases with distance from the Sun – at the distance
of Jupiter (about 5 AU) the illumination is only about 1/25 that at Earth. Unmanned
satellites presently en route or planned to the vicinity of Jupiter and beyond thus move
into a region in which the zodiacal light is much less bright than it appears to the
terrestrial observer. Thus a second step in the decontamination of the photometric
data is within the technology of the 20th century.

    To free the observations from the photometric contributions due to the starlight
and the associated diffuse galactic light calls for such a gigantic leap that we are

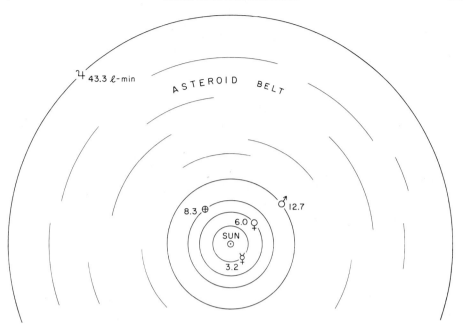

Fig. 7-3.   Portion of solar system within the orbit of the planet Jupiter. Distances are shown in *l*-min.

showing it in three successive illustrations, each on a linear scale in itself but separated from each other by factor jumps in scale.

Figure 7-4 shows the distribution of known stars and star systems within 17 *l*-yr of the Sun. The solar system proper is a small dot on this scale – the orbit of Pluto (5.6 *l*-hr) being about 10^{-4} the distance to the nearest drawn circle (5 *l*-yr). The solar gravitational domain extends outward until a neighboring star's gravitation exceeds that of the Sun. This domain probably contains a myriad of objects that are too small to be self-luminous and not sufficiently numerous to result in absorption or scattering that would produce observable effects in the night sky (see Chapter 6). The stars plotted in Figure 7-4 include ten that are brighter than $m = 6$ and thus are visible; these stars would remain individually visible to a traveler anywhere within the sphere of the plot. Further, the Sun would also become a visible star to add to the general sky – at a distance of 5 *l*-yr, the Sun would be a 0.7-mag. star (about as bright as Altair); at a distance of 10 *l*-yr, a 2.2-mag. star (Polaris); and at 15 *l*-yr, a 3.1-mag. star (a little brighter than ε Eridani, 3.7, and τ Ceti, 3.5). It is of passing interest that ε Eridani, at a distance of 10.7 *l*-yr, and τ Ceti (11.9 *l*-yr) are stars physically rather similar to the Sun and were prime candidates in a search for signals from possible intelligent civilizations in a program called, appropriately, OZMA.

Let us return to our purpose in showing Figure 7-4 – the search for a location at which the cosmic light might be isolated from the remaining foreground light from galactic stellar sources. As mentioned, the ten stars that are visible to terrestrial observers – plus the Sun – would remain as discrete stars. The light from the 34 fainter stars in the plot would total to the equivalent of 104 10th-mag. stars, which, if spread

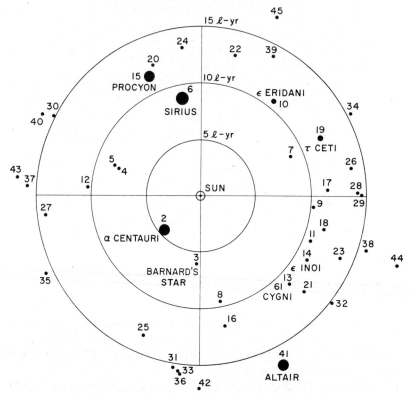

Fig. 7-4.   Known stars within 17 *l*-yr of the Sun (in a sphere, projected on a plane). The numbers correspond to the order of radial distance from the Sun. The angular position on the plot is the star's right ascension. Stars that appear close together on the plot are not necessarily close in space, since the third dimension – for example, the star's declination – cannot be indicated (taken from van de Kamp, 1971: 103, Table 1).

over the entire celestial sphere (41 253 sq deg), would contribute the equivalent of only 0.0025 10th-mag. star per sq deg – only a trivial fraction of the total starlight that concerns us in our cosmic search.

Thus an expedition to any spot in the local stellar domain would not be helpful – the Milky Way would still dominate the night sky. This conclusion is, of course, obvious with our present knowledge of our galactic system. After all, 15 *l*-yr is a trivial distance in terms of galactic dimensions, as shown in Figure 7-5, in which we note that the distance from the Sun to the galactic center is some 10 kpc (32 600 *l*-yr). With respect to the outer circle of Figure 7-5, with its radius of 15 kpc (49 000 *l*-yr), the outer circle of Figure 7-4 (15 *l*-yr) would reduce down to a circle smaller than the dot indicating the Sun.

It is obvious that the next step in our odyssey is to leave our Galaxy, which we do in Figure 7-6. Three of the objects shown in Figure 7-6 are visible – the two Magellanic clouds and the nebula in the constellation of Andromeda (Figure 7-7). The former are conspicuous, patch-like phenomena well-known to residents of the southern hemi-

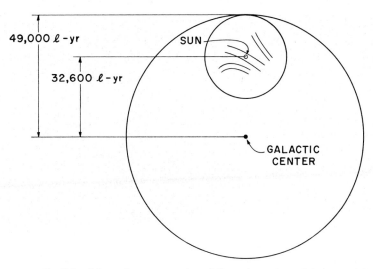

Fig. 7-5.   Schematic representation of dimensions of our Galaxy.

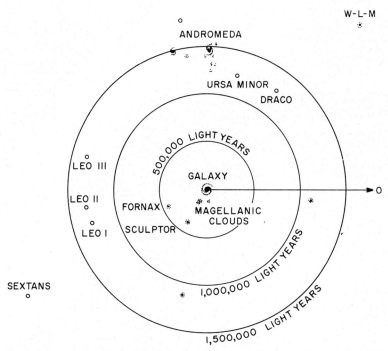

Fig. 7-6.   Nearby extragalactic objects plotted with our Galaxy in the central position.

sphere. The latter appears as a 4th-mag. object; its 'nebular' appearance is easily seen with a small telescope, or even with binoculars. Clearly, an observer located at random in the domain covered by Figure 7-6 is in the ideal location for the measurement of cosmic light. For example, a point at the apex of an equilateral triangle (libration

Fig. 7-7.   The extragalactic nebula in Andromeda.

point) equidistant from the Andromeda nebula and our Galaxy would reveal each of these two objects as of the 4th mag., and the rest of the sky would be only as bright as the totality of the light from the other and more distant galaxies.

Here we might interrupt to recall the dilemma posed by Olbers (see Chapter 2), with extragalactic objects replacing 'suns' in the argument. We may paraphrase Olbers (1826):

If there really be extragalactic nebulae in the whole of space, and to infinity, and if they are placed at equal distances from each other, or grouped into systems, their number must be infinite, and the whole vault of heaven should appear as bright as a typical extragalactic object; for every line which may be supposed to emanate from our eye towards the sky, would necessarily meet an extragalactic object, and thus every point of the sky would bring to us a ray of cosmic light.

Under these assumptions, the cosmic light should present us with a surface brightness equal to that of a typical object such as the Andromeda nebula. But, according to the integration of de Vaucouleurs, the cosmic light is some 40 times fainter. The full significance of this fact is confused by the systematic red shift of the light of the extragalactic objects.

Before making our final leap, note the large jump in the scales between Figures 7-5 and 7-6 – from the 49 000 l-yr for the larger circle in Figure 7-5 we make a more-than-10-fold jump in going to the smallest circle of Figure 7-6. As in all of our transitional steps, the earlier domain is dwarfed by the succeeding one.

A rough idea of the nature of the problem of the cosmic light may be appreciated by a study of Figure 7-8, in which we again change the scale by a large factor, this time

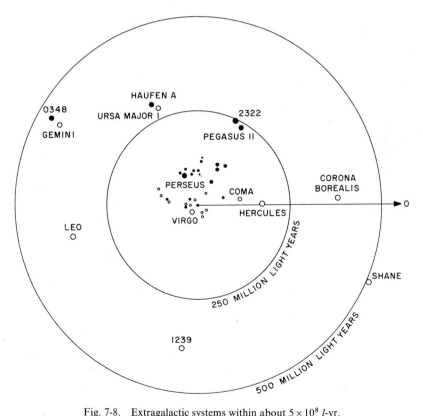

Fig. 7-8.   Extragalactic systems within about 5×10^8 l-yr.

by about 300, stretching the imagination as we contemplate the fact that the light from many of the plotted objects originated at a time when our Galaxy was about two revolutions back (period of revolution of our Galaxy is about 2×10^8 yr), and millions of years before the appearance of even primitive astronomers on our planet. By no means are all of the extragalactic objects within a sphere of 5×10^8 yr shown in Figure 7-8. For example, there are severe observational limitations in obtaining a complete count – the circle of the Milky Way of our Galaxy constitutes a 'zone of avoidance' as far as extragalactic objects are concerned because of the interposition of the galactic dust concentrated in and near that plane. Observation of the extragalactic domain becomes more and more effective as the observer looks progressively away from the Milky Way and towards the galactic pole – here is where an accurate photometric measurement of the cosmic light could be helpful to the cosmologist.

We have called attention to the drastic scalar jumps made as we moved from the Earth-Moon system (Figure 7-2) to the solar system (Figure 7-3), to the nearby-star domain (Figure 7-4), and to galactic associations (Figure 7-8). Another figure that might have been added would have had as its outer arc the horizon of our knowledge of the universe – say 10^{10} l-yr – another scale change of 20 over that of Figure 7-8.

It should be mentioned that not all the figures have the same physical significance. Figure 7-2 involves the observer in the center of a sphere looking outward through the nightglow layer in logarithmic steps to the Moon's distance. Figure 7-3 is basically a two-dimensional plot, the plane of the page being essentially the plane of the ecliptic, in which are concentrated the planets and the zodiacal cloud. Figure 7-4 is again a sphere projected onto a plane, the observer being centrally located. In Figure 7-5 the plane of the page is the galactic plane. Figure 7-6 may be considered as an observer-centered sphere, even though the known extragalactic objects tend to be *observed* selectively away from the galactic plane as a result of the galactic dust and the resulting zone of avoidance. But in Figure 7-8 a new feature enters – the distant extragalactic objects suffer a spectroscopic red shift that increases with the distance to the objects. The interpretation of the red shift as a doppler shift – implying the outward movement of the distant nebulae with velocities increasing with distance – has dramatized the concept of an expanding universe. In this picture the observer is on the surface of a bubble that is steadily getting larger, so that an observer on the surface of this hypothetical bubble sees his neighbors systematically retreating from him. Thus the observer in our part of the Galaxy is not in a central position, as he might egotistically think from the apparent symmetry of the repulsive motions away from him!

We have now explored the region from which the cosmic light originates via the dramatic vehicle of the imagination, with the subterfuge that we might find a spot for making the desired measurement. Several illustrations back we had actually far exceeded the physical possibility of such an observational mission, except under the supervision of a science-fiction enthusiast, for whom 'the impossible takes only a little longer'. So we practical writers must return, albeit reluctantly, to our small blue planet (number 3) ruled by a suburban star that is circulating around the inner city of a Galaxy which, in turn, is a member of a family of similar aggregations.

As we have seen, the elimination of the light from the nightglow has already been accomplished aboard both unmanned and manned satellites during the first 14 yr of the space age. Furthermore, present satellite programs involving expeditions to and beyond Jupiter promise the reduction of the zodiacal light to low levels of brightness. Thus it is likely that, by the end of the first few decades of the space age, measurements of the cosmic light will become available. The reduction of the integrated starlight suggested by the three entries in our earlier table is illustrated in Figure 7-1, in which the contribution from integrated starlight to the night-sky light is shown contrasted with that from cosmic light based on de Vaucouleurs' integration. It is obvious that a large potential gain is made available by going to fainter stars for the start of the integration. This approach calls for the use of a large telescope, for which faint stars appear individually rather than as part of the background, leaving the diffuse galactic light as the final photometric contaminant to an observer in the galactic system.

The prognosis for a successful assault on the cosmic-light measurement problem is thus clear – observe toward the galactic pole with a large telescope on a satellite in the outer part of the solar system – hint to astronomers who may now inhabit one of the satellites of Jupiter or Saturn!

In spite of the difficulty in disentangling the several components of the light of the night sky, and especially of extracting the cosmic light, we wish to emphasize our felicitous astronomical location in that no individual component of that light predominates so completely that all of the others are lost in the glare of the one. It is possible to imagine circumstances such that the sky would hide Nature's secrets rather than reveal them. For example, the zodiacal cloud may have been much more dense during early periods of our solar system (and of course might become more dense in the future), so dense that the zodiacal light would be so bright as to restrict or even preclude sidereal, galactic, and extragalactic research. The Milky Way, under such circumstances, would be so overwhelmed photometrically as to take away from us one of the aesthetically beautiful features of our planetary life. The same Milky Way loss would have occurred if the OH nightglow had been concentrated in the visual part of the spectrum rather than in the IR, in which the human eye is insensitive. Without the visibility of the Milky Way, would William Herschel have been inspired to gage the stars, or J. C. Kapteyn to set up the system of systematic star counts that led to our understanding of galactic structure?

These comments are made to temper our disappointment that we have to admit that the integrated cosmic light has been inadequately measured. We have only recently partly escaped physically from our terrestrial prison into nearby space and have tried in this book a more adventurous foray by way of outward glimpses through our atmospheric prison walls.

References

de Vaucouleurs, G.: 1949, *Ann. Astrophys.* **12**, 162.
Olbers, W.: 1826, *Edinburgh New Phil. J.* **1**, 141.
Roach, F. E. and Smith, L. L.: 1968, *Geophys. J., Royal Astronomical Soc.* **15**, 227.
van de Kamp, P.: 1971, *Ann. Rev. Astron. Astrophys.* **9**, 103.

EPILOGUE

Finally, we are impressed with the fact that a glance at the night sky is an adventure in both space and time. The cosmos is not so much at our fingertips as at our eyeballs – the multiple aspects of the universe present themselves freely to us every night. The incredibly varied sources of energy in the universe emit light quanta that impinge on our tiny retinal nerve endings after traversing both spaces and time periods so vast that we can visualize them, in a crude fashion, only with the help of numbers that are traumatically large.

A single visual sweep of the night sky brings us all sorts of light – light that left the nightglow zone, a mere 100 km distant, only about 1 ms ago (Figure E-1). Light that left our nearest neighboring star, α Centauri, about 4 yr ago; light that left the brilliant

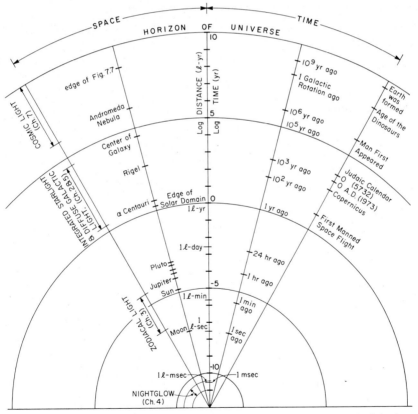

Fig. E-1.   Schematic plot of the Universe on logarithmic scales. Distances are shown in light-time units; corresponding look-back times are indicated.

star Rigel about the time of Copernicus. Light that left one of our neighboring galaxies, the beautiful Andromeda nebula, about the time Man first appeared on Earth. And light that left some of the most distant nebulae – say 3000 Mpc (10^{10} l-yr) distant, at about the time Earth was formed.

The reader interested in the possibility that astronomers may now be seeing 'to the edge of the world' in both space and time might profitably read Sandage (1972). In that paper (and earlier ones in the series) there is a critical discussion of the relationship between the observed spectroscopic red shift and the distance-time of the object under scrutiny.

The light we see represents its source at the time the light started on its journey. What has happened to that source in the time since? What has happened to the Andromeda nebula in the million years since the light we see tonight left there? And how about Rigel, whose light left only 500 yr ago – a fleeting moment in astronomical time?

We have the whole of past time before our wondering eyes at night – and the old and the young quanta arrive together to give an illusion of simultaneity. Sorting out the illusions and fastening onto the elusive realities is one of the most fascinating of puzzles – and it may start with a cursory glance at the night sky.

We urge the reader to put down this book and take such a glance – on a clear night, you can see forever!

Reference

Sandage, A.: 1972, *Astrophys. J.* **178**, 25.

SUBJECT INDEX

GEOPHYSICS AND ASTROPHYSICS MONOGRAPHS

AN INTERNATIONAL SERIES OF FUNDAMENTAL TEXTBOOKS

1. R. Grant Athay, *Radiation Transport in Spectral Lines.* 1972, XIII + 263 pp.
2. J. Coulomb, *Sea Floor Spreading and Continental Drift.* 1972, X + 184 pp.
3. G. T. Csanady, *Turbulent Diffusion in the Environment.* 1973, XII + 248 pp.

Forthcoming:

5. R. Grant Athay, *The Solar Chromosphere and Corona*
6. J. Iribarne and W. Godson, *Atmospheric Thermodynamics*
7. Z. Kopal, *The Moon in the Post-Apollo Era*
8. Z. Švestka and L. De Feiter, *Solar High Energy Photon and Particle Emission*
9. A. Vallance Jones, *The Aurora*
10. R. Newell, *Global Distribution of Atmospheric Constituents*
11. G. Haerendel, *Magnetospheric Processes*

Date Due

Due	Returned	Due	Returned
FEB 9 1979			
APR 1 1979			
JUL 21 1979			
SEP 2 5 1979			
FEB 1 9 82			
APR 23 82			
DEC 1 8 1984			
JUL 1 0 1986			
OCT 1 87			
APR 0 1 1988			
OCT 0 1 1989			
APR 0 1 1990	FEB 02 1990		
MAR 0 2 1990	FEB 13 1990		
APR 0 1 1990	MAR 3 0 1996		
OCT 0 1 1996	SEP 2 9 1999		
IN PHYSICS READING RM		IN PHYSICS READING RM	
IN PHYSICS READING RM		IN PHYSICS READING RM	

Continental Entertaining for the Young and Busy

ALSO BY ALINA ŻERAŃSKA

The Art of Polish Cooking

ALINA ŻERAŃSKA

Continental Entertaining for the Young and Busy

DAVID McKAY COMPANY, INC.

NEW YORK

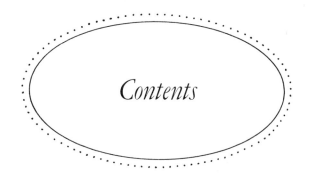

Contents

*Recipes followed by an asterisk (*)
can be found by
consulting the Index*

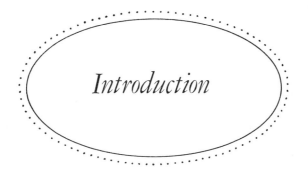

Introduction

"Amici mores noveris,
non oderis."

HORACE

(Learn the customs of your friend,
and there will be no hate.)

Food is one of the joys of life. Should we allow mass production and modern packaging to kill what generations have cherished and found so valuable? The ancient Chinese philosophers believed that if a man tries to count on his finger tips those things in life that truly give him enjoyment, he will invariably find that food is in first place. His interest in the delights of the table lasts till the end of his days, even after all others have long vanished. A good meal was as much the highlight of every cave man's gathering as it was in the enlightened era of Socrates. An elaborate dinner preceded the presentation of Shakespearian plays at King James' court.

It has taken centuries of experimenting to perfect the skill of proper seasoning to bring out the best in every kind of food. Most of us can broil a steak, but it takes imagination to create something delectable out of simple carrots. It is the use of herbs, spices, and dressings that make a dish unique. To know how to utilize them—that is the real art of cooking. And at the same time it is the simplest thing in the world, if you have reliable recipes with exact measurements, such as those presented in this book.

Today we can take advantage of the knowledge of the past and derive freely from the culinary wisdom of various nations. We can take what suits us, tossing away whatever we find obsolete, overly time-consuming, too complicated, or too heavy for our contemporary tastes and needs. The quality of products used is an important factor, but no less so than the choice of dishes, which should be easy enough for a modern woman to prepare without help in the kitchen and without toiling there for hours.

The recipes in this book were selected for their simplicity —but they are *not simplified* or *Americanized.* They are all authentic, but in my method of preparation I take full advantage of our American blenders, mixers, and certain prepared ingredients that make the work so much easier. It is Continental cooking at its best with all the variety of such party dishes as mushroom pies, fish baked in cream, or crisp canapés, special snacks for guests, elegant pastries flavored with rum and vodka, and delicate desserts sprinkled with fragrant cordials. These recipes offer diversity, yet they contain no ingredients difficult to find in America. This is original Continental cooking that requires no lengthy preparations.

The book is arranged under categories that make it easy to choose the right dishes for the right occasions. You will find refreshingly original ideas for breakfasts when you are entertaining house guests, for intimate luncheons, coffee klatches, evening parties, and late, late snacks. You can combine Continental recipes from several countries into one meal most suitable for your needs. You may serve a French hors d'oeuvre, a Swiss entrée, and a Hungarian dessert, or choose one nationality as a theme for your party.

Monotony takes the spark out of everyday living. We need diversity in the art of cooking. No wonder that in cosmopolitan metropolises like New York or San Francisco one finds restaurants offering foods from every corner of the world. No wonder that every good supermarket in the United States presents a selection of foreign delicacies. But unhappily, one can so often predict exactly what will be served at a cocktail or dinner party. The ubiquitous chips and dips will certainly be there, as well as the tossed salad and braised beef chunks over noodles or rice under the false pretense of being the famous "Boeuf Stroganoff." And the invariable pies and cakes will surely follow.

Fortunately, an increasing number of young hostesses with inquisitive minds are seeking to cull from the rich heritage brought to this country from overseas. They feel that investigating the unknown is very exciting.

Traveling through Europe with the help of this book, you will sneak into a Swedish kitchen to watch Mushroom Sauce simmering on the stove. The next moment you will see what is cooking in Norway. You can have a look into some famous Danish restaurants, or sail across the Baltic sea to taste a Polish fruit soup. Next you may find yourself climbing over the Carpathian Mountains for a traditional Czech Christmas with goose and cabbage, or swinging into Budapest for Chicken Paprika and delicious tortes. And why not end the day in a Viennese garden sipping coffee with ice cream?

The French and Jewish cuisines have made the greatest impact on the cookery of most European nations. This is probably so because food has occupied an especially important place in both the French and Jewish ways of life. The result has been a great inventiveness and artistry in the preparation of traditional dishes developed during the long history of these peoples. In the case of Jewish cookery, certain dishes were also influenced by religious commandments and the necessity of making the most out of whatever was available. Settling in many countries, the Jews brought with them new, delectable, and economical ideas for preparation of foods. The French, on the other hand, being born gourmets, created a belief that French dishes were especially elegant, and indeed everyone has wished to imitate their cuisine. Those two influences intermingling with one another and with the national cookery of European countries have brought most savory results.

Continental cooking is different from the kind of food preparation we usually find in America. There is more emphasis on soups, which are served in Europe every day in every home, and on individual treatment of each kind of vegetable. There is more inventiveness in ways of serving less expensive cuts of meats, and endless possibilities for fish—a real delight for weight and budget watchers. Different seasonings and different herbs are used. You notice a much greater variety of breads and pastries.

Quite often Continental cuisine scares our young and active people. Many cookbooks are available that have been written by connoisseurs of fancy foods served in internationally famous restaurants, or by graduates of famous cooking schools that provide chefs for famous hotels. Those books offer recipes for tasty dishes but they are too elaborate to be practical for ordinary people. No housewife in Paris, Vienna, Budapest, or Copenhagen cooks that way.

My recipes are not restaurant recipes, but old family recipes tested and improved through decades. They come from a variety of Europeans I have encountered during the years I spent in different countries, during my travels, and through my international contacts. They are classic, traditional, but simple enough to be used by busy people in their homes. The original dishes presented here are for imaginative, creative individuals with a curiosity for something new, for those who crave food prepared and served with Old World elegance, but at the same time with respect for our modern need for simplicity.

Breakfast
with
Company

PREFACE

The critical period in any marriage is breakfast time. Very few of us wake up in good humor. Not many have the time or appetite for an elaborate meal before rushing to work. But on Sundays and holidays, on carefree vacation mornings, eating alone or with friends who have stayed overnight, breakfast can become a very special occasion.

After a night out or a party, when everyone is still a little sleepy and tired, the first meal should not be heavy or elaborate. None of the recipes in the following chapter is complicated or requires more than 15 minutes for preparation and cooking.

· 5 ·

To prevent breakage during cooking, eggs should be taken from the refrigerator and brought to room temperature the first thing in the morning, or even the night before. Placing a few rolls in the oven for five minutes takes little effort, but makes them crisp and appetizing—a pleasant diversion from our usual toast. A touch of originality is also a good idea. Eggs Benedict, for example. The sauce may be prepared the day before, or a jar of a commercial one will do. The favorite Danish cheese sandwich doesn't even require cooking. Omelets are real problem-solvers for a busy hostess. It doesn't even take much invention to prepare a different kind of omelet each time, because any leftover vegetable or meat, separate or combined, that you find in your refrigerator can make a perfect omelet filling.

We can draw endless ideas from the richness of Continental cooking. The kind of breakfast people are used to depends greatly on the climate and traditions of their native land. A Frenchman who likes to dine late and lavishly is satisfied with a morning croissant and a cup of coffee or cocoa. The mild Italian climate calls for little more. But in northern Europe, it is different. Centuries ago, shepherds in the rugged Scottish mountains used to eat a bowl of hot porridge, and today descendants of the Vikings on Swedish hills or Danish farms have preserved the old culinary traditions and like to start their day the same way.

An Englishman wakes up with tea. He takes his time about dress, building his appetite for a hearty meal that consists of our traditional eggs and bacon, or of the all-time British favorites—kippers, kidneys, or even a small steak. This is followed with muffins or toast and marmalade.

A Ukrainian peasant enjoys the nourishing warmth of milk soup with noodles or barley before he goes out to his fields. Poles prefer a light, but more refined breakfast, consisting of a variety of dishes. They may start with a soft-boiled egg, then try a hard roll with butter and ham, or farmer's cheese, sipping some strong coffee with a lot of milk, nipping on a few radishes generously covered with butter, then finishing up with another roll and honey. In July, on both sides of the Carpathian Mountains, wild strawberries with cream make a special breakfast treat.

. . . .

For Breakfast Menus

1. Eggs Vienna Style
2. Jewish Eggs with Matzos
3. Scrambled Eggs from Debrecen
4. Eggs à la Bonaparte
5. French Eggs Benedict
6. Stuffed Eggs from Warsaw
7. French Plain Omelet
8. French Stuffed Omelet
9. French Cheese Omelet
10. French Omelet on Toast
11. Flea-Market Omelet
12. Omelet For a Crowd, from Lichtenstein
13. Ham Omelet from Sweden
14. French Strawberry Omelet
15. Hungarian Mushrooms with Eggs
16. Spinach Pancakes from Finland
17. Noodle Pancakes from Portugal
18. Swiss Fried Cheese
19. Danish Cheese Sandwich

. . . .

Eggs Vienna Style

The famous English writer Oscar Wilde thought that an egg is always an adventure: it may be prepared in so many different ways. And indeed, boiled eggs do not always have to be just plain boiled eggs. It all depends on how they are seasoned and served, or in what mood you eat your breakfast. A pleasant festive atmosphere is easily created with imagination and a careful setting of the table.

8 eggs
Salt, pepper, paprika
2 teaspoons butter

If you take the eggs directly from the refrigerator, place them in a bowl with warm water for 5 minutes to warm them up and prevent breakage during cooking.

Heat water in a large saucepan. Carefully place the eggs in enough boiling water to cover them, using a tablespoon. Cook on high heat for 3 minutes.

Remove the eggs from the saucepan. Break the shells, and, using a teaspoon, transfer egg whites and yolks into 4 sherbet glasses or dessert bowls. Sprinkle with salt, pepper, and paprika. On the top of each portion place ½ teaspoon butter. Work fast to prevent cooling.

SERVES 4.

Jewish Eggs with Matzos

The old father in *Fiddler on the Roof* stresses in a melodious way the importance of traditions in giving each individual a feeling of security and his own place in life. Attachment to tradition gives a special spark and unity to Jewish family life.

Jewish dishes smell tantalizing, taste delicious, and look attractive. Here is one, a Passover dish, with the traditional matzos, sure to please your guests of any background, not just your Jewish mother-in-law.

1 large onion, chopped
3 tablespoons butter
2 matzos, 7 × 7, in small pieces
6 eggs, well beaten
⅛ teaspoon pepper
Salt to taste

Fry the onions in hot butter until golden.

Rinse the matzos with boiling water in a colander just to moisten lightly. Mix with the eggs, season with salt and pepper. Add the egg mixture to the onions and fry over moderate heat until set.

SERVES 6.

. . . .

Scrambled Eggs from Debrecen

Debrecen is a small city in Eastern Hungary with a fine old university (1588) and plenty of tradition. Debrecen happens also to be famous for its sausage; it is hard to match this smoked delicacy anywhere in Central Europe. We cannot duplicate it in America, but we can have scrambled eggs Debrecen style, using a good-quality Polish sausage or summer sausage.

4 ounces bacon, diced
½ pound Polish sausage, peeled and sliced
3 green peppers, cored and sliced
2 tomatoes, cut into wedges
10 eggs, beaten
Salt
4 ounces grated Parmesan cheese

Fry the bacon until crisp. Add sausage, peppers, and tomatoes. Fry until peppers are tender, stirring.

Add the eggs and cook at medium heat, stirring slowly; season with salt. When done, sprinkle with cheese and serve.

SERVES 6.

· ·

Eggs à la Bonaparte

Napoleon was more interested in war than in food, but still he had some favorite dishes. As French people eat very little breakfast, he liked his eggs later in the day. Eggs, served Napoleon's favorite way, make a nice brunch. Hot rolls are a good accompaniment.

3 medium tomatoes
6 eggs
2 lettuce leaves, finely shredded
Salt and pepper
1 teaspoon lemon juice
6 teaspoons Mayonnaise*
12 fillets of anchovy
12 slices of dill pickle
6 lettuce leaves

*The asterisk indicates that there is a separate recipe for the item elsewhere in this book. To locate the recipe, please consult the Index.

Cut the tomatoes in half. Remove the central pulp of the tomato halves and turn upside down to drain.

If taking eggs directly from the refrigerator, place them in warm water for 5 minutes to prevent breakage during cooking. Transfer them to boiling water, cover, bring to a boil again, and cook for 4 minutes. Remove from the pot.

Place some shredded lettuce inside tomato halves. Sprinkle with salt, pepper, and lemon juice.

Shell the eggs carefully, without breaking the egg whites as the egg yolks are soft. Place each whole egg inside a tomato half. Place one teaspoon Mayonnaise over each egg and cross it with two anchovy fillets. Place 2 slices of pickle alongside each egg.

Arrange lettuce leaves on 6 plates. Place stuffed tomatoes on lettuce.

SERVES 6.

French Eggs Benedict

This is an excellent choice for all morning gatherings, shower parties, family reunions, or just a special treat for out-of-town visiting friends. Cook your sauce a day ahead and reheat carefully in a double boiler. Never bring it to boil. A commercial imported French sauce may be used also to save labor.

8 lettuce leaves
8 slices toast
8 slices ham
8 strips bacon, cut into halves, fried crisp
8 poached or fried eggs
8 tablespoons Sauce Béarnaise* or Sauce Hollandaise*
¼ cup green parsley,† finely chopped

Place lettuce leaves on 8 plates. Arrange toast on lettuce, cover with ham, place bacon over it, and the eggs on the bacon. Place one tablespoon sauce on each egg. Sprinkle with parsley.
Work fast to keep the toast and the eggs warm.

SERVES 8.

. . . .

Stuffed Eggs from Warsaw

Polish stuffed eggs make an attractive breakfast that even a child can prepare. It is a good idea to boil and chop the appropriate ingredients and stuff the shells during a free moment the day before. Refrigerate them well covered, then reheat in a few minutes in a frying pan when your guests come to the kitchen hungry.

---

†I specify green parsley in order to avoid confusion with the white parsley's root, which is widely used abroad.

6 large eggs
3 tablespoons bread crumbs
⅓ cup sour cream
1 ½ tablespoons green parsley or dill weed
2 tablespoons green onion, finely chopped
Salt and pepper
4 tablespoons butter

Place the eggs in tepid water for 15 minutes. Transfer into a kettle, cover with cold water, bring to a boil. Cook for 5 minutes. Place in cold water to cool.

Cut the eggs lengthwise, using a sharp knife. Do not crush the shells. Scoop out the eggs and chop finely. Combine with 1 tablespoon bread crumbs, sour cream, and herbs. Season. Return the mixture to the shells. Cover with the remaining bread crumbs, flatten with a knife.

Heat the butter in a large skillet. Fry the eggs, stuffing side down, till golden brown.

SERVES 6.

French Plain Omelet

Omelets are the answer for young and inexperienced cooks who want to serve something different and tasty, but do not yet have the skill or patience for complicated and time-consuming dishes. You can never go wrong with this recipe if you will just remember a few hints: Never use more than 6 eggs for one 9 to 10 inch frying pan. Do not beat your eggs in a mixer or blender; they should not be entirely blended. And do not let the omelet set completely on the top. It should be moist.

6 eggs
2 tablespoons cold water
Salt and pepper to taste
1 tablespoon butter

Beat the eggs with water, salt, and pepper, giving them 15 strokes with a fork. Pour into skillet over melted butter. Fry on medium heat. When the eggs begin to set, lift the edges of the omelet with a knife, and allow some of the runny egg to flow under, repeating at intervals. When most of the egg is set, fold the omelet in two with the help of a large knife, and transfer onto a serving platter.

SERVES 3.

French Stuffed Omelet

With one basic recipe, you can make an endless variety of French omelets. Look for ideas in your pantry and in your refrigerator. Almost anything will do as a filling: leftover vegetables (peas, asparagus, broccoli, cauliflower), chopped cold cuts, or chopped leftover meats.

6 eggs
2 tablespoons cold water
1 tablespoon chopped green parsley or dill weed
Salt and pepper
1 tablespoon butter
2 cups stuffing

Beat the eggs with water, herbs, salt, and pepper, giving them 15 strokes with a fork. Pour into a skillet over melted butter. Fry on medium heat. When the eggs begin to set, lift the edges of the omelet with a knife and allow some of the runny egg to flow under, repeating at intervals.

Heat the filling. When most of the egg is set, spread the filling on the omelet, and fold it in two with the help of a large knife. Transfer onto a serving platter.

SERVES 4.

· ·

French Cheese Omelet

It is not true that French cuisine is extremely elaborate. Perhaps the chefs from Cordon Bleu cooking schools have time for complicated procedures, but the average French woman does not. French wives, mothers, and single girls often work for their living, seldom have household help, and cannot spend endless hours in the kitchen. Their cooking is simple, but highly imaginative.

6 eggs
2 tablespoons cold water
1 tablespoon green parsley, chopped
Salt and pepper
1 tablespoon butter
¾ cup grated or shredded cheese

Beat the eggs with water, parsley, salt, and pepper, giving them 15 strokes with a fork. Pour into a skillet over melted butter. Fry on medium heat. When the eggs begin to set, sprinkle with the cheese. Lift the edges of the omelet with a knife and allow some of the runny egg to flow under, repeating at intervals. When most of the egg is set, fold the omelet in two with the help of a large knife. Transfer onto a serving platter.

SERVES 3.

. . . .

French Omelet on Toast

My sister-in-law lived in France for many years. This is how she learned to serve an omelet to her guests in the morning.

1 French Plain Omelet*
6 slices toast
2 tablespoons butter
6 slices tomatoes, grilled
1 tablespoon green parsley, chopped

Prepare a plain omelet, but do not fold it. Cut into 6 wedges.

Spread the toast with butter. Place a piece of the omelet on each piece of toast, top with tomato slices, sprinkle with parsley.

SERVES 6.

. . . .

Flea-Market Omelet

There is some truth to the notion that Parisians are perverse. Since 1222 they have flocked during Holy Week to the annual Scrap Iron and Ham Fair, originated by King Philip Augustus as a test of the strength of the spirit. The idea was to expose the faithful to extreme temptation, so that their Lenten fast would do their souls more good. Pork butchers displayed their meats in front of the Cathedral of Notre Dame; anyone who gave in and ate was punished by both the church and the state. Peddlers took advantage of the gathering by selling various goods, and this is how flea markets started. Booths were also set up to feed the hungry in a way approved

by the church. This is one of the tasty Lenten dishes served at
the Ham Fair, which you can serve at any season.

6 eggs
2 tablespoons cold water
1 tablespoon dill weed
Salt and pepper
1 ½ cups boiled potatoes, thinly sliced
1 tablespoon butter

Beat the eggs with water, dill, salt, and pepper, giving
them 15 strokes with a fork. Fold in potatoes, stirring gently.
Pour into a skillet over melted butter. Fry on medium heat.
When the eggs begin to set, lift the edges of the omelet with
a knife and allow some of the runny egg to flow under, repeating at intervals. When most of the egg is set, cut into four and
serve.

SERVES 4.

· ·

Omelet for a Crowd, from Lichtenstein

The biggest problem with omelets is that it is impossible
to cook one with more than 6 eggs at a time unless you use
several frying pans. People in Lichtenstein, a tiny country in
the Rhine valley, have a more practical solution. This is a
perfect omelet for feeding a larger number of guests, as for
example when you are having a family reunion.

Cook your potatoes a day ahead in their skins.

¼ pound butter
1 tablespoon oil
6 medium potatoes (1 ½ pounds), cooked, peeled, sliced
½ teaspoon garlic salt
12 slices bacon, fried crisp, crumbled
½ pound mushrooms, finely sliced
12 eggs, blended
Salt and pepper to taste

Heat half the butter and the tablespoon of oil. Fry potatoes until golden. When almost done, sprinkle with garlic salt. Place in a square baking dish, sprinkle with bacon.

Sauté mushrooms in half the remaining butter for 5 minutes.

Season the eggs with salt and pepper, combine with mushrooms. Pour over potatoes. Place in a 475° oven for 5 minutes. Dot with the remaining butter and bake for 10 more minutes in 350° oven until set.

SERVES 10.

. . . .

Ham Omelet from Sweden

Here is another good way of using up leftover ham that takes very little work and is always successful. A simple, elegant breakfast for the man she loves can give any woman the self-assurance of a blond Swedish nymph.

2 strips bacon, sliced
1 ½ cups chopped ham
3 eggs
½ cup coffee cream

Arrange the bacon in a 9″ round baking dish and place under broiler until bacon is transparent. Add ham and broil until it is lightly golden.

Beat the eggs and cream with a fork for 2 minutes. Pour over the ham. Bake in 400° oven until set, in about 7 minutes.

SERVES 2.

- -

French Strawberry Omelet

Here is an easy dish, delicate and light as a feather, elegant as the Old World. A pleasant meal depends on tasty food, but also on the manner in which it is served and the general outlook of our surroundings. As many of us often eat our breakfast in the kitchen these days, it is important to keep the place tidy and attractive. The time required for baking this omelet is enough to permit brewing good, strong coffee, and washing or putting away all the dishes required for preparing the omelet.

One 10-ounce box frozen strawberries, defrosted
4 eggs, separated
Pinch of salt
2 tablespoons sugar
4 tablespoons flour
½ teaspoon baking powder
1 teaspoon butter

The night before, take the strawberries out of the freezer before going to bed.

IN THE MORNING: Drain off the syrup from the strawberries and save. Beat the egg whites with salt until stiff but not dry. Beat the egg yolks with sugar for 3 minutes. Mix the flour with baking powder. Sift over the egg yolks, alternately adding egg whites in small portions. Mix lightly.

Melt the butter in a 9-inch round baking dish and spread evenly over the bottom. Transfer the egg mixture into the pan. Distribute strawberries evenly. Bake in 400° oven for 15 minutes. Serve immediately, with a pitcher of strawberry syrup on the side.

SERVES 3.

Hungarian Mushrooms with Eggs

This dish makes a perfect late breakfast or brunch that takes only 15 minutes to prepare.
Serve with toast and butter.

¼ pound butter
2 onions, sliced
1 ½ pounds mushrooms, sliced
Salt and pepper
1 teaspoon green parsley, chopped
1 cup sour cream
½ cup flour
4 eggs
4 parsley sprigs

Set aside 1 tablespoon butter for the eggs.

Heat the remaining butter in a large skillet, add the onions. Fry for 3 minutes. Add the mushrooms, fry for 5 minutes. Season with salt, pepper and green parsley. Mix the flour with sour cream and some juice from the skillet. Add to the mushrooms, mix, bring to boil. Fry the eggs sunny side up.

Place each egg on a separate plate, spoon mushrooms around each. Decorate with parsley sprigs.

SERVES 4.

. . . .

Spinach Pancakes from Finland

Everyone in Finland looks forward to Midsummer Festival celebrations. People travel to the countryside just as we do on the 4th of July. Only instead of our fireworks, Finns build huge bonfires. They love to sit, dance, and sing around the sparkling flames. The whole country turns into one big party. The festival is an occasion for meeting new friends, starting new romances, and ending the old ones.

All kinds of food are prepared in large quantities; spinach pancakes are served throughout the day.

2 ⅔ cups milk
2 teaspoons salt
2 cups flour
2 tablespoons butter, melted
2 eggs
1 teaspoon sugar
½ pound fresh spinach leaves, finely chopped
½ cup salad oil
1 can cranberry sauce

Beat the milk with salt, flour, and butter in a blender for 2 minutes. Let stand for ½ hour. Add eggs and sugar, beat some more. Combine with spinach.

Heat two 6-inch skillets and brush well with oil. On both skillets at the same time, fry a number of individual thin pan-

cakes on both sides over medium high heat until golden
brown.

Serve cranberry sauce in a separate bowl.

SERVES 6.

. . . .

Noodle Pancakes from Portugal

These quick pancakes, popular in Portugal, are very
much to the American taste. Nourishing and simple, they
make an ideal late-morning summer breakfast that is even
adaptable to cooking while camping out. It is practical to cook
noodles the previous night, or to use leftovers.

Serve with tomato salad and radishes.

4 eggs
Salt and pepper
4 ounces grated cheese
2 cups cooked noodles
2 tablespoons butter

Beat the eggs until foamy, season with salt and pepper.
Fold in the cheese, mixing slightly. Fold in noodles, mixing
gently.

Fry small pancakes in hot butter on both sides until
golden.

SERVES 4.

. . . .

. . . .

Swiss Fried Cheese

Cheese makes a perfect breakfast food. The Swiss often start their day that way. Serve fried cheese on toast and with freshly brewed strong coffee.

8 slices, 3″ by 1½″ by ⅓″ thick, Swiss or Gruyère
   cheese
½ cup milk
½ cup flour
¼ pound butter

Dip each slice of cheese in milk; roll in flour. Repeat three times. Fry in hot butter on both sides until golden. Serve immediately.

SERVES 4.

. . . .

Danish Cheese Sandwich

Some people feel that a monarchy is not a totally dated or stupid idea. They think that a dynasty is a permanent symbol of the continuity of the state that evokes more respect than an elective office. And the monarchy does not have to be as costly as it is in Great Britain. The late Danish King Frederik IX lived and behaved like an ordinary citizen. He liked to mingle with his people, taking strolls or bicycle rides on the streets of Copenhagen. He also favored simple food, like this open sandwich, believing that protein for breakfast is of fundamental value.

12 radishes, chopped in a blender
1 cup cottage cheese
6 slices French bread
12 slices Swiss cheese, rolled
½ cup green onion, chopped

Combine the radishes with cottage cheese. Spread on the bread. Arrange rolls of Swiss cheese on each sandwich. Sprinkle with green onions.

SERVES 6.

*Elegant Luncheons
for an
Intimate Group*

PREFACE

Every successful hostess has mastered two fundamental arts: the art of planning and the art of sensible organization.

Planning means choosing dishes most appropriate for a well-balanced meal. You need to take several factors into consideration: the time of day, the number of guests, an attractive combination of flavors and colors, and the feasibility of food preparation that will keep you in the kitchen as little as possible during the party or even during that day.

Nothing spoils the mood of a social gathering more than a tired, tense hostess. And it doesn't make sense to invite friends when instead of enjoying their company you must tend the pots all the time.

At midday a light simple meal is preferable, cold on summer days, warm in winter. Most of the recipes from the Evening Snacks chapter could be very well used for a luncheon, but not vice versa—for an obvious reason. The hostess is apt to be at home in the morning, and she has time for preparation.

Many of the following dishes selected for luncheons are informal in character, perfect for family reunions or entertaining close friends. In this category are all the casseroles, noodles, dumplings, or pancakes. To sophisticated out-of-towners, serve the Swedish Seafood Sandwich, especially if you do not wish to be bothered with cooking. The best way to impress a few gourmet cooks is to serve them a soufflé.

The recipes in this chapter include suggestions for how to plan your luncheon menu around a particular dish. Some of these recipes require cutting, grating, or frying; but they are meant for a small gathering and so will not take much time to prepare. The nicest luncheon parties are those for small groups of 6 to 8 at the table. That way you can have a general conversation in an informal, intimate atmosphere.

Most of the lovely desserts for your luncheon menus may be prepared a day ahead. They may also be served on various other occasions, especially after dinner or for an evening snack. Many of the coffee cakes and cookies from other chapters are also good at a noon meal, but tortes meant for a dinner or an evening buffet would be too heavy at this time of day.

It is practical to take a piece of paper and write down your menu, and beside each dish the amount of time needed for cooking and at what time you should start to prepare it. That way you will see immediately how much work is left for the day of the party, especially for the last hour. Then decide on the exact time at which you want to serve your meal; in that way you will know at what time each pot should go on the stove or into the oven. Let your kitchen timer alert you to this. It is so easy to lose track of the time when you are with your favorite people.

Invite your guests to arrive half an hour before luncheon, one or one and a quarter hours before you want to serve dinner. If one of your guests is tardy, his time of relaxation before the meal will be cut short. If he has not arrived by the time you are ready to call your guests to the table—wait, but no longer than 10 minutes, unless there is a snowstorm or other exceptional excuse for his delay. Cooking is an art and should be treated with due respect. You haven't gone through all these careful preparations only to have everything ruined because someone started shaving too late.

If you want to be relaxed and in good spirits when entertaining your guests, do as much of your shopping as you can one or two days ahead. Sleep longer, spend enough time making yourself pretty. Choose a becoming outfit. Even a luncheon gathering calls for an elegant if not dressy outfit. The latest fad for wearing highly informal clothing or dowdy sweaters for every occasion certainly does not add much to womanly charms.

Set the table, prepare serving platters, pots, and pans for your luncheon in the morning. And when your guests are already there, believe me, it does not matter whether the draperies and windows were cleaned that same week or if the rug could have used additional vacuuming. No one notices details but the hostess. It is more sensible anyway to clean up after the party than before, unless your home is really dirty.

And no one will know if a dish did not exactly turn out as you expected unless you tell. The best cooks have kitchen mishaps, but those who gain fame for their skills have the wisdom of silence.

For Luncheon Menus

ENTREES

1. Cheese Pie from France
2. Shrimp Pie from Belgium
3. Potato Casserole from Strasbourg
4. Brussels Sprouts Belgian Casserole
5. Sausage Casserole from Venice
6. Pepper and Rice Hungarian Casserole
7. Roman Rice with Cheese and Tomatoes
8. Italian Baked Zucchini
9. Turkish Noodles and Yogurt
10. Hungarian Stuffed Peppers
11. Bulgarian Peppers Stuffed with Cheese
12. Cheese Dumplings from Slovakia
13. Swedish Seafood Sandwiches
14. Russian Pancakes
15. Swiss Cheese Soufflé
16. French Salmon Soufflé

SWEET DISHES AND DESSERTS

17. French Pancakes
18. Jewish Cheese Pancakes
19. Jewish Apple Pancakes
20. Scandinavian Pancake Torte
21. Cherry Dumplings from Carlsbad
22. Rice and Apple Pudding from Slovakia
23. Swedish Rice Pudding
24. Apple Soufflé from Vienna
25. Stuffed Apples from Finland
26. Viennese Apple Mousse
27. Apple Compote from Poland

28. *Norwegian White Veils*
29. *Viennese Chestnut Purée*
30. *Bavarian Mocha Delight*
31. *Strawberry Delight from Old Russia*
32. *Strawberry Pie from Belgium*
33. *Strawberries in Wine Sauce from Vienna*
34. *Cream of Rhubarb from Finland*
35. *Finnish Cranberry Cream*
36. *Danish Cranberry Dessert*
37. *Apricot Dessert from Prague*
38. *Christmas Fruit Dessert from France*
39. *Almond Cheese from Russia*
40. *Chocolate Cheese from Russia*
41. *Raspberry Cheese from Russia*

. . . .

Cheese Pie from France

One of the greatest of all French inventions is *Quiche Lorraine,* a cheese pie that conquered the world faster and more permanently than Napoleon ever dreamed of.
Serve it fresh and warm from the oven with a bowl of salad in Sauce Vinaigrette,* asparagus in butter dressing, French bread, sweet butter, and well-chilled white wine.

1 unbaked 9-inch pie shell
6 slices bacon
1 ¼ cups Gruyère or Swiss cheese, cut into small slices
2 cups light cream (half'n half)
3 eggs
½ teaspoon salt
¼ teaspoon nutmeg
⅛ teaspoon pepper

Fry the bacon until crisp, place on absorbent tissue to cool. Crumble the bacon evenly over the bottom of the pie shell. Sprinkle the cheese over the bacon to cover.
Blend the cream well with eggs and spices. Pour into the pie shell over the cheese. Bake in 400° oven for 15 minutes. Reduce the heat to 325° and bake 30 minutes longer.

SERVES 6.

. . . .

Shrimp Pie from Belgium

This is a Belgian variation of the French *Quiche,* ideal for seafood lovers. I had it once in a small restaurant overlooking the ocean where the shrimp were fresh and cooked just right. It was served with buttered Brussels sprouts and tomato salad.

1 unbaked 9-inch pie shell
1 cup cooked small shrimp
2 tablespoons butter
¼ cup dry white vermouth
1¼ cups Swiss cheese, cut into small slices
2 cups light cream
3 eggs
½ teaspoon salt
¼ teaspoon nutmeg
⅛ teaspoon pepper

Place the pie shell in 400° oven for 5 minutes. Cool.

Sauté shrimp in butter for 3 minutes, add vermouth, and cook at high heat, stirring until all liquid evaporates. Distribute shrimp evenly in the pie shell. Sprinkle with cheese.

Blend the cream with eggs and spices. Pour it over the cheese. Bake in 400° oven for 15 minutes. Reduce the heat to 325° and bake 30 minutes longer.

SERVES 6.

. . . .

Potato Casserole from Strasbourg

Germans are known to be great potato eaters. One can hardly imagine a dinner in a German home without potatoes on the table. The German influence in Strasbourg makes potatoes an important part of the daily menu, but the French population of that lovely border city always tries to make an inspired dish out of potatoes.

This tasty casserole makes a nourishing luncheon, and is

a good way to use leftover ham or sausage. Serve it with a big bowl of lettuce.

6 medium potatoes, thinly sliced
1 ½ cups ham, chopped
1 ½ cups yellow cheese, chopped
2 eggs
1 tablespoon tomato purée
1 cup strong bouillon from cubes
1 teaspoon paprika

Place ⅓ of the potatoes in a well-buttered round baking dish. Cover with half of the ham, sprinkle with ⅓ of the cheese. Add a layer of half the remaining potatoes, then a layer of the rest of the ham. Sprinkle with half the remaining cheese; then cover with the rest of the potatoes.

Mix the eggs with tomato purée, bouillon, and paprika. Pour over the potatoes. Sprinkle with the rest of the cheese. Place in a hot 450° oven for 45 minutes.

SERVES 6.

· ·

Brussels Sprouts Belgian Casserole

This is a quick family luncheon for people who do not eat eggs for breakfast. You use up your leftovers and serve a tasty, wholesome one-dish meal in no time. If necessary, you may substitute any leftover vegetable for Brussels sprouts. Serve with tomatoes and radishes.

2 cups sliced cooked potatoes
2 cups cooked Brussels sprouts
2 eggs
1 cup milk
Salt and pepper

Combine potatoes with Brussels sprouts and distribute evenly in a buttered 9-inch glass pie dish. Combine the eggs with milk, salt, and pepper in a blender. Pour over the vegetables. Bake at 350° for 20 minutes.

SERVES 6.

Sausage Casserole from Venice

This nice, light luncheon should be served with warm garlic bread, plenty of sweet butter, and, of course, a bottle of Chianti. Soft Italian music in the background will make the day perfect.

2 pounds zucchini
1 ½ cups sausage, sliced
1 cup bread crumbs
1 cup Parmesan cheese, grated
Dash of garlic
1 teaspoon salt
¼ teaspoon pepper
2 tablespoons parsley, chopped
2 eggs, separated
1 teaspoon paprika

Cut off the ends of the zucchini and quarter the squash. Cover with boiling salted water and cook for 15 minutes; drain. Chop fine, combine with sausage and bread crumbs mixed with half of the cheese, garlic, salt, pepper, and parsley. Mix well with egg yolks. Fold in whipped egg whites. Place in a well-buttered 9″ × 12″ baking dish. Sprinkle with the remaining cheese mixed with paprika. Bake in 350° oven for 45 minutes.

SERVES 6.

· ·

Pepper and Rice Hungarian Casserole

Green pepper is a rich source of vitamins and a tasty vegetable which can be used in a great variety of dishes. This convenient luncheon, a whole meal in one pot, is excellent for cold winter days. Hearty eaters may like to add a few wieners and warm French bread to the menu.

½ pound lean bacon, chopped
7 medium onions, sliced
1 cup rice (do not use precooked rice)
1 ½ cups bouillon from cubes
3 large green peppers, seeds removed, chopped
One 1-pound can tomatoes
Salt to taste
¼ teaspoon pepper
1 teaspoon paprika

Fry the bacon in a deep, 9-inch pan until pale golden. Drain half of the fat. Add the onions and fry till transparent, but not brown. Add the rice and fry for 5 minutes. Add the bouillon, green peppers, and tomatoes with juice. Season with salt, pepper, and paprika. Cover and cook at medium heat for 25 minutes.

SERVES 6.

· ·

Roman Rice with Cheese and Tomatoes

Somehow cheese and tomatoes always find a way into an Italian dish. I often wonder what Italian cuisine was like before it adopted tomatoes. And that was not so long ago. Tomatoes were first cultivated in South America. When they were brought to Europe, people used the lovely red fruit just for decoration. An Italian cookbook from the beginning of this century recommends cooking tomatoes for at least two hours because they were supposed to be so difficult to digest.

Serve Roman rice with lettuce and Italian dressing.

1 large onion, chopped
2 tablespoons butter
1 cup rice
2 cups strong bouillon from cubes
½ pound pizza cheese, shredded
¼ cup bread crumbs
2 tablespoons butter
2 large tomatoes, cut into thick slices
3 tablespoons butter

Sauté onions in hot butter until golden, add rice, and fry for 2 minutes. Add bouillon, cover, and simmer until rice is tender. Place ⅓ of rice in a buttered baking dish, sprinkle with half of the cheese, cover with half of remaining rice, then repeat layers of cheese and rice. Sprinkle with bread crumbs, dot with butter. Place on a lower shelf under broiler until golden.

Sauté tomatoes in hot butter on both sides, place on the rice, and serve.

SERVES 4.

· ·

Italian Baked Zucchini

Italians have interesting ways of cooking vegetables. The climate in Italy is perfect for cultivating them and they play an important part in Italian menus. In fact, most vegetables popular in Europe today were introduced at one time or another into various countries by visiting Italian princes and artists. Serve baked zucchini with baked sausage and a tomato salad.

3 medium zucchini
3 cups bread crumbs
1 large onion, grated
3 tablespoons green parsley, finely chopped
1 cup Parmesan cheese, grated
Salt and pepper
1 egg, beaten
3 tablespoons butter
1 tablespoon paprika

Parboil zucchini in salted water for 5 minutes. Cut each one in half lengthwise and remove the pulp.

Chop the pulp, add bread crumbs, onions, parsley, half of the cheese, salt and pepper. Combine with the egg.

Fill the zucchini shells, dot with butter, sprinkle with the remaining cheese and paprika. Bake at 350° for 30 minutes in a well-oiled baking pan.

SERVES 6.

Turkish Noodles and Yogurt

My best friend lived for several years in Istanbul with a Turkish family. She knows all the secrets of harems and many special tricks Turkish women use to please their men. One of them is this quick, economical Turkish recipe.

The secret of this dish is to serve the noodles very hot and the yogurt very cold. Be sure your yogurt is really fresh and plain, without any added flavoring.

4 cups cooked noodles
3 tablespoons margarine, melted
2 medium onions, finely chopped
2 tablespoons margarine
½ pound mushrooms, sliced
1 pound ground beef
Salt and pepper
2 cups yogurt

Mix noodles with melted margarine. Fry the onions in hot margarine until pale golden, add the mushrooms, fry 3 more minutes. Mix in the meat and fry, stirring constantly until the meat is crisp and brown. Season well with salt and pepper. Mix with the noodles. Warm the luncheon plates.

Beat very cold yogurt. Place in a chilled bowl and refrigerate.

Transfer hot noodles onto individual plates. Offer each guest some yogurt to pour over his noodles.

SERVES 6.

· ·

Hungarian Stuffed Peppers

Hungary has exceptionally fertile fields. Abundant rain and sunshine provide favorable conditions for growing vegetables. Hungarian green peppers and paprika have a special rich flavor due to the soil and climate. Pepper pods are small and sweet. In the green stage they are used as a vegetable. When ripe they acquire a bright orange-red color; they are hung up to dry, and then ground and turned into paprika.

This ever-present vegetable and spice is a surprisingly new acquisition. It was introduced by the Turks during the invasion, but won popularity slowly. Only during the last hundred years has paprika become widely used in Hungarian cuisine. Today, if you visit a Hungarian home, you may be sure you will be offered something with green peppers or paprika. And Hungarian cooks do not just sprinkle food with paprika, they use it by spoonfuls.

1 large onion, chopped
1 tablespoon bacon drippings
1 ½ pounds ground pork
1 cup cooked rice (do not use instant rice)
1 egg
1 tablespoon paprika
1 teaspoon salt
⅛ teaspoon pepper
8 green peppers, cored

SAUCE
2 cups tomato purée
1 tablespoon instant flour
1 teaspoon sugar
Salt to taste

Fry the onions in hot drippings until golden. Mix with the meat, rice, and egg. Season with paprika, salt, and pepper.

Stuff the peppers loosely with this mixture. Arrange in a baking dish. Combine the ingredients of the sauce. Pour over stuffed peppers. Bake in 375° oven for 1 hour.

SERVES 8.

. . . .

Bulgarian Peppers Stuffed with Cheese

During the summer we do not feel much like cooking or eating anything heavy. Bulgarians get their proteins in warm weather from yogurt and cheese rather than from meat dishes. Stuffed peppers are often served to vacationers in seashore hotels at the popular Bulgarian resort whose name is "Zlatni Pyassatsi," meaning "Golden Sands."

Serve with rice, tomato salad, and yogurt or buttermilk.

4 green peppers, seeds removed (peppers are stuffed whole)
½ pound cheese, American or Gouda, diced
½ cup milk
1 large onion, finely chopped
1 tablespoon butter
2 hard-boiled eggs, finely chopped
2 chicken bouillon cubes
1 cup water, boiling
2 tablespoons tomato purée

Cook peppers in a kettle with boiling water for 2 minutes; drain. Melt the cheese in milk on medium heat, stirring. Do not boil. Remove from heat.

Sauté the onions in butter until lightly golden. Add to cheese. Combine with eggs. Stuff the peppers. Place them in saucepan. Dissolve bouillon cubes in boiling water; add tomato purée. Pour over the peppers. Cover and simmer for 25 minutes.

SERVES 4.

. . . .

Cheese Dumplings from Slovakia

This is a nourishing tasty dish for a family luncheon, especially handy when you entertain relatives with children, and you are tired or in a hurry. These cheese dumplings take only 10 minutes to prepare, and everyone is fed and pleased. Serve with a green salad.

2 pounds farmer cheese
4 eggs
½ teaspoon salt
⅔ cup cream of wheat
1 ½ cups flour
4 tablespoons butter
3 tablespoons bread crumbs

Press the cheese through a strainer. Mix with eggs and salt. Add cream of wheat and flour. Mix with a spoon for a few minutes.

Drop small pieces of the dough from a teaspoon into a kettle with boiling, salted water. Bring to a boil, cook at high heat for 3 minutes. Transfer with a slotted spoon to a warmed serving dish. Melt the butter over low heat. Add the bread crumbs, fry for a few minutes until golden, stirring. Pour over dumplings.

SERVES 6.

Swedish Seafood Sandwiches

Do you like shower parties in the morning? They are simpler, easier on the hostess than an evening shower, and a

much less elaborate meal may be offered. Invite your guests
for 11 a.m., and around noon serve these delicious sandwiches
with chilled rosé wine, followed by an elegant dessert.

12 slices pumpernickel
¼ pound sweet butter, soft
3 heads butter lettuce
24 jumbo shrimp, cooked
12 ounces canned crab meat
1 cup Mayonnaise*
⅓ cup dill weed, fresh or dry, coarsely chopped

Spread the bread generously with butter. Cover with a
thick layer of lettuce leaves. Arrange 2 shrimp and a tablespoon of crab meat on the lettuce. Garnish with Mayonnaise.
Sprinkle with dill.

SERVES 12.

. . . .

Russian Pancakes

Each nationality has its own variety of pancakes. The
famous French *Crêpes Suzette* are strictly a dessert. Russians
serve more hearty pancakes, which can make a meal in themselves. *Blini,* as these are called, are garnished with many
nourishing things, which should be served in separate dishes.
Each person takes one or several kinds with each pancake.
Most commonly used as a garnish are chopped hard-boiled
eggs, melted butter, sour cream, sardines, anchovies, or—if
you can afford it—caviar. Buckwheat flour is available in Jewish or health food stores.

¾ ounce yeast
1 teaspoon sugar
3 cups milk
1 ¾ cups all-purpose flour, sifted
1 ¾ cups buckwheat flour, sifted
2 eggs, separated
2 tablespoons butter, melted
¼ teaspoon salt
¼ pound shortening

Combine the yeast with sugar and 1 cup milk. Stir in the all-purpose flour. Cover and place in an oven heated to 100° until doubled in size. Combine with buckwheat flour and the rest of the milk to make a thick batter. Beat well. Gradually beat in the egg yolks, butter, and salt.

Beat the egg whites until stiff but not dry, and fold into the batter. Cover and let stand in the oven until doubled in size.

Ladle about ¼ inch thick of the mixture onto a hot greased griddle or skillet to make 5-inch pancakes. Cook until golden on the bottom, turn and lightly brown other side. Keep warm while cooking the remaining batter.

SERVES 8.

Swiss Cheese Soufflé

To make a soufflé is the easiest thing in the world. It just consists of a white sauce, a purée of something with an attractive flavor, and well-whipped egg whites.

Use eggs at room temperature, and separate them carefully. Use clean, dry containers and tools. One drop of egg yolk, moisture, or grease will prevent your egg whites from whipping high and stiff—the important factor in a successful soufflé.

Ask your guests to come to the table before you take the soufflé from the oven.

3 tablespoons butter
1 cup coarsely grated Swiss cheese
3 tablespoons instantized flour
1 cup milk
3 eggs at room temperature
Dash pepper
Pinch nutmeg
½ teaspoon salt
⅓ teaspoon cream of tartar

Spread a deep 1½-quart casserole with 1 tablespoon butter and sprinkle with 1 tablespoon cheese.

Melt the remaining butter. Stir in the flour and cook for 2 minutes without browning. Stir in the milk and cook until thick, stirring steadily. Remove from heat. Add egg yolks, one at a time, stirring them briskly into the sauce; add pepper and nutmeg.

Whip the egg whites with salt until stiff. Add cream of tartar, and beat some more. Stir 2 tablespoons of the egg whites into the sauce. Stir in the remaining cheese, saving one spoonful. Fold in the remaining egg whites. Mix gently. Turn the mixture into the baking dish. Sprinkle with last spoonful of cheese. Bake in 350° oven for 50 minutes. Serve immediately.

SERVES 4.

. . . .

French Salmon Soufflé

Here is a delicacy which at first glance seems elaborate, requiring special skills, but once you have tried it, you know that it is indeed quite simple to prepare. Serve salmon soufflé with tossed salad in Sauce Vinaigrette,* warm French rolls, sweet butter, and a bottle of good white wine.

3 tablespoons onions, finely chopped
3 tablespoons sweet butter
3 tablespoons instantized flour
¾ cup milk
Juice from canned salmon
1 ½ tablespoons tomato paste
4 egg yolks
¾ cup canned salmon, finely chopped
½ cup grated Parmesan cheese
Pinch pepper
1 teaspoon parsley, finely chopped
5 egg whites
½ teaspoon salt
¼ teaspoon cream of tartar

Sauté the onions in hot butter, add the flour, and fry for a few minutes. Add milk and juice from salmon. Cook at low heat, stirring until thick. Stir in tomato paste. Remove from heat and stir in egg yolks one by one, then the salmon, and half of the cheese. Season with pepper and parsley.

Whip the egg whites with salt until stiff. Add tartar and beat some more. Stir 2 tablespoons of the egg whites into the salmon mixture. Fold in the remaining egg whites. Mix gently. Turn into a deep 1½-quart baking dish which has been buttered and sprinkled with cheese. Sprinkle the remaining cheese over the top. Bake in 375° oven for 35 minutes.

SERVES 5.

· ·

French Pancakes

Not many people today can eat pancakes for dessert on top of a full meal as these *Crêpes Suzette* were originally supposed to be served. But they can nicely supplement a light salad luncheon.

Ask your guests to be on time. Prepare your French pancakes just before their arrival. Skip the cocktails and ask everyone to come to the table right away. Serve Sauterne wine.

BATTER
4 eggs
2¼ cups flour
1 teaspoon sugar
½ teaspoon vanilla extract
2 cups milk
4 tablespoons sweet butter, melted
1 tablespoon cognac or curaçao

ORANGE SPREAD
4 tablespoons sweet butter, soft
⅓ cup sugar
⅓ cup curaçao

3 tablespoons butter for frying, melted

IN THE MORNING: Beat the eggs with flour, sugar, and vanilla for a few minutes. Add milk, butter and cognac; beat some more. Let stand several hours.

30 MINUTES BEFORE LUNCH: To make the spread, cream the butter with sugar, adding curaçao in a thin stream.

Heat an 8-inch skillet. Brush lightly with butter. Pour ¼ cup of the batter and tilt the pan immediately so that the batter will spread over one entire bottom of the pan. Cook the pancake on both sides. Spread lightly, fold, and place in a warmed dish. Repeat until all the batter is used. Stack the pancakes in a covered dish in a warmed oven. Serve as soon as possible.

SERVES 8.

- -

Jewish Cheese Pancakes

Jewish pancakes, or *blintzes,* are famous in America. They may be served with various stuffings, but cottage cheese is most often used. *Blintzes* make a lovely luncheon. They may be prepared and filled in the morning, refrigerated, and fried just before serving. For a well-balanced meal, serve with a big bowl of lettuce.

BATTER
2 eggs
1 ½ cups milk
1 cup flour
½ teaspoon salt

4 tablespoons butter for frying

FILLING
1 pound creamed cottage cheese
¼ cup sugar
2 egg yolks
¼ teaspoon cinnamon
2 tablespoons sour cream

SPREAD
1 cup sour cream

Combine the batter in a mixer. In a separate bowl mix the filling well.

Heat a 6-inch skillet. Brush lightly with butter. Pour in ¼ cup of the batter and tilt the pan immediately so that the batter will spread over entire bottom of the pan. Cook over low heat on one side only until the top of the pancake is dry. Stack the pancakes on a plate.

Place a tablespoon of the filling in the center of each pancake on the fried side. Fold to form envelopes. Fry in butter until golden on both sides. Serve hot with sour cream in a separate jar.

SERVES 6.

. . . .

Jewish Apple Pancakes

If you want to be original, choose apple *blintzes* for your luncheon dessert. They are especially good following a tasty crabmeat or turkey salad. Cook the pancakes in the morning, but it is best to mix the filling, to fill the pancakes, and fry them again just before serving.

Prepare pancakes as for Jewish Cheese Pancakes*

APPLE FILLING
2 cups apples, coarsely grated
2 tablespoons almonds, finely chopped
3 tablespoons confectioners' sugar
½ teaspoon cinnamon
1 egg white, whipped

FOR FRYING:
3 tablespoons butter

FOR SPRINKLING:
3 tablespoons sugar

Mix the apples with almonds, sugar, and cinnamon. Fold in the egg white. Place some of the filling in the center of each pancake on the fried side. Fold to form envelopes.

Fry immediately in butter until golden on both sides. Sprinkle with sugar and serve.

SERVES 6.

Scandinavian Pancake Torte

Blueberries are plentiful in Scandinavian forests. In this northern section of Europe, where most of the fruit has to be imported, blueberries are pure delight during the summer months. They are used in many delicious ways—this one is especially popular.

2 eggs
1 cup milk
½ cup water
1 cup flour
¼ teaspoon salt
4 tablespoons butter
1 pint blueberries
½ cup sugar
1 cup whipping cream
2 tablespoons confectioners' sugar

Combine the eggs, milk, water, flour, and salt in a mixer. Heat an 8-inch frying pan spread with butter. Fry pancakes on medium heat on one side only. When half done, sprinkle with blueberries mixed with sugar and fry until the top of the pancake is set. Pile pancakes on a serving platter one over another. Cut pancake torte into wedges.

In a separate bowl serve cream whipped with sugar.

SERVES 6.

· ·

Cherry Dumplings from Carlsbad

The most important crowned heads of Europe used to come to the famous Czech spa of Carlsbad. The old emperor of Austria-Hungary, Franz Joseph, strongly believed in the miraculous powers of its mineral waters. The leading hotel in Carlsbad, the Hotel Pupp, once regarded as perhaps the most elegant spa hostelry in Europe, frequently offered its famous guests Czech national dishes, among them incomparable fruit dumplings. I like them best with cherries, but they are equally popular filled with small prune plums, apricots, and blueberries.

Serve a small seafood hors d'oeuvre, then cherry dumplings. You won't need a dessert.

DOUGH
2 tablespoons butter, soft
¼ teaspoon salt
1 egg
¼ cup farmer cheese
2 ¼ cups instantized flour
½ cup milk

FILLING
1 pound cherries

TOPPING
¾ cup bread crumbs
4 tablespoons butter
½ cup sugar

Beat the butter until creamy. Add salt and the egg; beat for 5 more minutes. Add cheese, flour, and enough milk for a soft, elastic dough. Roll out on a floured board, cut into small squares. Place one cherry on each square. Roll into dumplings in your palms.

Cook in a large kettle with salted boiling water for 7 minutes. Transfer with a slotted spoon into a warmed platter.

Fry bread crumbs in hot butter for a few minutes. Pour over the dumplings. Sprinkle with sugar.

SERVES 8.

. . . .

Rice and Apple Pudding from Slovakia

This simple dish will please people with a sweet tooth. It was always a children's favorite, especially welcome after a morning on sleds or a good snowball fight.

4 cups peeled, cored, and shredded apples
1 tablespoon cinnamon
½ cup sugar
4 cups cooked rice (do not use precooked rice)
1½ cups sour cream
4 tablespoons confectioners' sugar

Combine the apples with cinnamon and sugar.

Arrange the rice and the apples in alternate layers in a well-buttered round baking dish. Make three layers of rice and two layers of apples. Cover and place in a 375° oven for 45 minutes.

Mix sour cream with confectioners' sugar. Serve in a separate dish.

SERVES 6.

· ·

Swedish Rice Pudding

In Sweden the Christmas season begins on December 13 with the Festival of Lights. Traditionally, on that morning in every home the eldest daughter gets up before daybreak to serve breakfast in bed to every member of the family. She lights seven candles on a crown and places it on her head, then makes the round of bedrooms, bringing coffee and buns on a tray covered with a beautifully embroidered white cloth. This is a signal that soon the darkness of the long Swedish winter will be gone and the sun will shine again. On that same day a snowy, sweet dessert is often served for lunch.

3 cups milk
2 tablespoons sugar
¼ teaspoon almond extract
1 cup rice (not precooked)
1 ¼ cups whipping cream
2 tablespoons confectioners' sugar
½ teaspoon vanilla extract
Strawberry preserves

Bring milk with sugar and almond extract to a boil; add rice and stir. Simmer well covered for ½ hour. Refrigerate for several hours.

Whip the cream; add sugar and vanilla. Combine with the rice. Place in dessert bowls and garnish with preserves.

SERVES 8.

. . . .

Apple Soufflé from Vienna

In my junior year of high school, I took German in addition to the obligatory Latin and French for the pure pleasure of understanding the dialogues and songs in Austrian movies. My teacher was a young Viennese girl who supplemented the principles of grammar by showing me how to make her mother's favorite soufflé.

1 ½ tablespoons butter
⅓ cup sugar
3 eggs at room temperature
2 tablespoons lemon juice
One 15-ounce jar of applesauce
Pinch salt
½ cup bread crumbs
One orange rind, grated

Cream the butter, adding sugar by spoonfuls, egg yolks one by one, and the lemon juice. Beat 5 minutes more. Combine with the apple sauce.

Whip the egg whites with salt until stiff.

Add egg whites to the applesauce mixture in small portions, alternating with bread crumbs combined with orange rind. Mix gently.

Turn into a deep 1½-quart casserole which has been buttered and sprinkled with bread crumbs. Bake in 375° oven for 35 minutes.

SERVES 4.

. . . .

Stuffed Apples from Finland

Finland is a small northern nation with a stubborn people and an orderly way of life. Finns are hard working and thrifty, but they also appreciate good company and good cooking.

Not many fruit trees can stand long Finnish winters. Apple trees are the sturdiest, and the Finns have learned to make the most out of them.

Stuffed apples make an elegant, light dessert requiring very little preparation.

3 tablespoons honey
3 tablespoons chopped walnuts
3 tablespoons graham-cracker crumbs
2 tablespoons orange liqueur (curaçao)
6 medium apples
2 teaspoons butter

Mix honey with walnuts, cracker crumbs and liqueur.

Cut the tops off the apples. Remove the seeds. Fill the holes with the stuffing. Dot with butter. Cover with the tops.

Place close to one another in a baking dish. Bake in 375° oven for 1 hour. Serve warm.

SERVES 6.

. . . .

Viennese Apple Mousse

Desserts are probably the most exciting and original element in the whole range of Viennese cookery. This one is particularly useful when, after baking a cake or making a creamy spread, we don't want to waste leftover egg whites.

The preparation of this dessert takes 10 minutes.

3 large egg whites
1 envelope plain gelatin
3 tablespoons cold water
1 tablespoon boiling water
4 tablespoons sugar
½ teaspoon almond extract
One 1-pound jar applesauce
¼ cup almonds, chopped

Whip the egg whites for 5 minutes.

Soak the gelatin in cold water for 5 minutes. Add boiling water. Place the cup with gelatin in a pot with hot water and stir until dissolved.

Add sugar to the whipped egg whites by spoonfuls, beating constantly. Add almond extract and gelatin in a thin stream, and beat some more. Fold in applesauce. Mix gently.

Place apple mousse in six sherbet glasses. Sprinkle with almonds. Refrigerate for several hours before serving.

SERVES 6.

. . . .

Apple Compote from Poland

An increasing number of people are beginning to complain about the chemicals added to packaged foods. This favorite Polish dessert is simple to prepare and pure as nature itself. It is also different. We have canned fruit of many kinds in our American supermarkets, but we cannot buy canned apple compote.

4 cups water
1 ½ cups sugar
1 teaspoon cinnamon
1 teaspoon lemon rind, grated
8 apples, peeled, cored, quartered
1 tablespoon lemon juice

Bring water with sugar, cinnamon, and lemon rind to a boil. Add the apples. Simmer 15 minutes. Add lemon juice. Chill thoroughly.

SERVES 8.

. . . .

Norwegian White Veils

This dessert has an amusing name, *"Tilslørte Bondepiker,"* which means "Peasants in Veils." The dish itself is not only tasty, but also very decorative. If you use some imagination, it may look like heads in white bonnets, as worn by peasant women in Norway.

⅔ cup sugar
3 tablespoons water
2 ½ cups bread crumbs
1 cup whipping cream
½ teaspoon vanilla extract
½ cup confectioners' sugar
2 cups applesauce

Dissolve sugar in water in a frying pan over medium heat. When the syrup browns, quickly stir in bread crumbs. Do not let the caramel solidify into hard pieces.

Whip the cream, add vanilla and sugar.

Spread half of the applesauce on the bottom of a 9-inch round glass pie dish. Sprinkle with ¼ of bread crumbs. Spread with ⅓ of the cream. Sprinkle with second quarter of bread crumbs. Spread with remaining applesauce. Sprinkle with half of the remaining bread crumbs. Scoop whipped cream over it, forming a circle of large balls. Sprinkle with the rest of the bread crumbs. Refrigerate for ½ hour.

SERVES 6.

· ·

Viennese Chestnut Purée

When I was a student in Paris, I loved to buy chestnuts roasted right on street corners over a bed of red coals. The tempting aroma of these small mobile stoves sometimes lured me to a chestnut stand several blocks away. It was customary then as now to eat them right from a paper bag while strolling on the boulevards. The hot shells burned the fingers, but freshly roasted warm chestnuts tasted heavenly on chilly winter days.

The French South, particularly the city of Grenoble, specializes in glazing chestnuts in hot syrup. The Viennese have always shared the French love of chestnuts, often serving them as a superb dessert—chestnut purée with whipped cream.

PUREE
2 pounds chestnuts
⅔ cup confectioners' sugar
1 teaspoon vanilla extract
2 tablespoons rum
3 tablespoons milk

CREAM
1 cup whipping cream
3 tablespoons confectioners' sugar
1 teaspoon vanilla extract

A DAY AHEAD: Cut each chestnut lengthwise half through. Soak in warm water for 25 minutes.

Heat the oven to 350°. Divide your chestnuts into 4 portions. Roast the first portion for 25 minutes. Immediately peel off and discard the shells and the brown inner skin while the second portion is roasting. Repeat until all chestnuts are roasted and shelled.

Place the chestnuts in a saucepan, cover with boiling water, cook for 20 minutes, and drain. Grind twice. Mix with sugar, vanilla, rum, and milk.

BEFORE SERVING: Whip the cream; add sugar and vanilla.

Place half of the cream in a serving dish. Strain the chestnuts over it. Decorate with the remaining cream around the sides of the dish.

SERVES 8.

. . . .

Bavarian Mocha Delight

This is a pure delight, not just because of its name, but because it will delight a busy hostess. It is as splendid as a fancy torte, but it does not require baking. Prepare everything early in the morning; it will take you only 20 minutes.

6 eggs
1 ⅓ cups sugar
2 tablespoons water
¾ pound sweet butter
2 tablespoons instant coffee
1 tablespoon vodka
1 pound lady fingers
¼ cup very strong coffee
Candied cherries

TO MAKE FILLING: Beat the eggs and sugar for 7 minutes. Add water, and heat until thick, still beating with hand mixer.

Cream the butter. Add the egg mixture to the butter by spoonfuls, beating all the time. Add the instant coffee, dissolved in vodka, in a thin stream; beat some more.

TO MAKE MOCHA DELIGHT: Line the bottom of a springform baking pan with ⅓ of the lady fingers. Sprinkle with half of the coffee. Spread with ⅓ of the filling. Cover with half of the remaining lady fingers, sprinkle with the rest of the coffee. Spread with half of the remaining filling. Cover with the rest of lady fingers, pushing them firmly into the filling. Refrigerate for 3 hours. Save the remaining filling at room temperature.

Remove the sides of the pan. Spread the remaining filling

on the top and sides of the dessert. Decorate with cherries. Refrigerate until serving time.

SERVES 8.

. . . .

Strawberry Delight from Old Russia

This is an old-fashioned dessert from the court of the Romanoffs, and the epoch of the last Russian tsar. In the happy days of his youth, the tsar's court was full of splendor, elegant people, and laughter. This strawberry delight was often served after sumptuous luncheons and dinners attended by Russian aristocrats and foreign ambassadors.

Serve it in your most elegant sherbet, champagne, or wine glasses.

2 quarts small strawberries, washed, hulled, and chilled
1 cup whipping cream
1 pint vanilla ice cream, softened at room temperature
¼ cup lemon juice
⅓ cup orange-flavored cordial

Whip the cream until stiff. Whip the ice cream in a mixer until fluffy. Fold in the whipped cream, lemon juice, and cordial. Fold in berries. Fill the glasses and serve immediately.

SERVES 8.

. . . .

Strawberry Pie from Belgium

How delicious rosy strawberry tarts are after a winter of heavy cakes! Take full advantage of strawberries when they are in season and at their best.

This simple recipe makes it possible to serve an attractive strawberry dessert without baking.

SHELL
1 cup graham-cracker crumbs
3 tablespoons confectioners' sugar
¼ pound butter, melted

FILLING
1 quart strawberries, washed and hulled
1 cup sugar
⅓ cup corn starch
1 tablespoon lemon juice

ON THE NIGHT BEFORE: Sprinkle strawberries with sugar and let stand overnight.

TO MAKE SHELL: Combine cracker crumbs with sugar. Stir in the butter. Save ⅓ of the mixture. Pack the rest of the mixture firmly into 9-inch pie pan and press to bottom and sides. Refrigerate for ½ hour.

TO MAKE FILLING: Drain off the juice from the strawberries and measure. Add enough water to make up 1¾ cups liquid. Blend corn starch with ¼ liquid. Slowly add the rest of the liquid, mixing. Cook on a low heat, stirring constantly until sauce boils and is clear. Place the saucepan into a larger one with boiling water, cover and cook for 15 minutes.

Remove from the heat, add lemon juice, and fold in strawberries. Cool until lukewarm. Pour into the pie pan. Sprinkle with the remaining crumb mixture. Refrigerate. For a more elegant dessert treat, top with whipped cream.

SERVES 6.

. . . .

Strawberries in Wine Sauce from Vienna

A simple bowl of strawberries may turn into an exciting dessert if you learn this secret from Emperor Franz Joseph's household. His subjects didn't know much about technology and their plumbing was not too great, but they knew how to dance, and they knew how to eat.

1 quart strawberries
6 egg yolks
⅔ cup sugar
1 cup white table wine

Wash and slice strawberries before your guests arrive.

JUST BEFORE SERVING: Rinse the mixing bowl with hot water to warm it up. Beat egg yolks with sugar in this bowl for 5 minutes. Heat the wine almost to the point of boiling, but do not boil. Add hot wine to the egg yolks in a thin stream, beating constantly. Beat for 5 more minutes. Place strawberries in serving bowls. Cover with wine sauce and serve.

SERVES 6.

. . . .

Cream of Rhubarb from Finland

In spite of the fact that Finland is such a small nation (less than 5 million people), and its land has been devastated by many wars, it is counted among the 12 countries enjoying the highest standards of living in the world. This is due in a great degree to the Finns' thrifty and hard working way of life. They try to make the most out of what they can grow in their severe climate. Rhubarb has an important place in the summer menus of every family.

2 cups rhubarb, cut into ½ inch pieces
¾ cup sugar
2 cups boiling water
3 tablespoons potato flour or corn starch
3 tablespoons cold water

Simmer rhubarb with sugar in boiling water for 15 minutes. Blend potato flour with cold water. Stir into stewed rhubarb; bring to a boil. Simmer for 3 minutes. Chill well before serving.

SERVES 6.

* *

Finnish Cranberry Cream

Nothing can build a healthy appetite better than a good session in a sauna. There are 350,000 rural saunas in Finland. They look like small log cabins and are heated by stoves filled with cobblestones. Water thrown on these hot stones produces steam. Sweating is stimulated by beating the body with birch twigs.

There is a legend that the Finns love to run out of the sauna naked, straight from the steam, and roll in the snow. If they do that, they don't like to admit it. My Finnish friends assure me, however, that even without this ancient shock treatment, sauna does wonders in rejuvenating the body.

1 cup boiling water
1 ½ cups cranberry juice
¼ cup instant cream of wheat
¼ cup sugar
Light cream

Add cranberry juice to boiling water, sprinkle in the cream of wheat, stirring. Cook at low heat for 7 minutes. Add sugar, cool a little, then beat until stiff. Pour into serving bowls and refrigerate until serving. Serve cream in a separate jar.

SERVES 6.

- -

Danish Cranberry Dessert

In our times of artificial coloring, flavoring, and hundreds of other chemical additives to our foods, many people turn to old-fashioned recipes, containing only natural ingredients. This Scandinavian old-timer, so simple to prepare, should please everyone who wants to eat naturally.

Cook in the morning or a day ahead. The consistency of this dessert should not be quite as thick as that of a pudding. Serve very cold with cream in a separate pitcher.

6 tablespoons potato starch or corn starch
½ cup cold water
1 ½ cups hot water
3 ¾ cups pure cranberry juice or other berry juice
⅔ cup sugar
1 cup coffee cream or whipped cream

Blend the potato or corn starch with cold water. Stir into hot water. Bring to a boil. Stir in the juice and half of the sugar. Bring to a boil, stirring, simmer for 2 minutes.

Pour into dessert bowls. Sprinkle with remaining sugar. Refrigerate until ready to serve.

SERVES 8.

· ·

Apricot Dessert from Prague

Apricots grow deliciously sweet on sunny slopes in Czechoslovakia. People in Prague like this simple, light, and healthful summer dessert which can be made with various other fruits and berries as well as apricots. Strawberries and raspberries may be used raw, but all others require cooking. With canned fruit readily available, we can serve this dessert in every season. Choose a creamy brand of cottage cheese that is not too salty.

One pound can apricots, drained
1 ½ cups creamed cottage cheese
2 tablespoons rum
3 tablespoons grated chocolate

Set aside 6 small slices of apricot for decoration. Press remaining apricots and the cottage cheese through a strainer. Add rum. Whip in a mixer until fluffy. Place in 6 sherbet glasses. Sprinkle with chocolate. Garnish with apricot slices. Chill before serving.

SERVES 6.

· ·

Christmas Fruit Dessert from France

Christmas in France is both a family holiday and a religious celebration. Each family arranges a manger, *la crèche,* in a prominent part of the house. This custom was introduced in Avignon by the family of Saint Francis of Assisi at the beginning of the 14th century, but did not become widespread in France until the 16th century. Children like to bring home rocks, branches, and moss for the manger setting. In Paris the little figurines used in the crèche are often original sculptures that are genuine works of art.

Having had a large meal after the midnight mass, French families eat lightly on Christmas day. Fruits are served in various ways, as in this elegant dessert.

12 ounce can cubed pineapple
12 ounce can pears
12 ounce can peaches
12 ounce can pitted black cherries
1½ cups creamed cottage cheese
¾ cup coffee cream

Drain the fruits. Set aside half of the cherries for decoration. Arrange fruits in a serving bowl. Whip the cheese with cream in a blender until smooth and fluffy. Pour over the fruit, sprinkle with remaining cherries, and serve.

SERVES 12.

. . . .

Almond Cheese from Russia

In all Central and Eastern Europe, Easter holidays are as important as Christmas. The great emphasis is on food, a real Easter feast that lasts for half a day. Traditional dishes are prepared for several days before the holiday. On Sunday morning they are displayed on a large table covered with a snowy white tablecloth and beautifully decorated with greenery.

Among the most famous Russian Easter specialties are tall, light and aromatic *babas* baked with dozens of egg yolks, and the cheese dessert, *pascha,* an easy-to-make, rich dessert, that is very different and tasty.

Prepare pascha a day ahead of your party. Serve it well chilled, cut into small slices.

1 pound creamed cottage cheese
5 large hard-boiled egg yolks
½ pound sweet butter, soft
2 ⅓ cups confectioners' sugar
1 teaspoon vanilla extract
½ teaspoon almond extract
¾ cup almonds, coarsely chopped
¼ cup candied orange rind, finely chopped

Rub the cheese with the egg yolks through a sieve. Mix.

Beat the butter until creamy, add sugar in small portions beating constantly. Add vanilla and almond extracts; beat for 5 more minutes.

Combine the cheese mixture with creamed butter, mixing thoroughly. Add almonds and orange rind; mix some more. Refrigerate for 1 hour.

Now form the mixture into a roll. Wrap it tightly in a cotton napkin which has been dipped in water and squeezed out. Refrigerate overnight.

BEFORE SERVING: Remove the napkin. Place the dessert on a serving platter, decorate with candied fruits and almond halves. Keep refrigerated until you are ready for it.

SERVES 8.

. . . .

Chocolate Cheese from Russia

The Russian cheese dessert, *pascha,* may be prepared with different flavors. Some people like it with raisins; others feel that it tastes best when made with chocolate.

In Russia special wooden molds with little holes for drainage are used for *pascha.* They imprint attractive designs on the cheese loaf. Some of the antique molds are real masterpieces of folk art. The wet napkin method lacks decorations, but produces satisfactory results.

1 pound creamed cottage cheese
5 large hard-boiled egg yolks
½ pound sweet butter, soft
2 cups confectioners' sugar
1 teaspoon vanilla extract
1 cup instant chocolate powder
¾ cup walnuts, coarsely chopped

Rub the cheese with the egg yolks through a sieve. Mix.
Beat the butter until creamy and add sugar in small portions, still beating. Add vanilla and beat for 5 more minutes. Combine with chocolate and the cheese mixture; mix some more. Add the walnuts. Refrigerate.

Now form mixture into a roll. Wrap it tightly in a cotton napkin which has been dipped in water and squeezed out. Refrigerate overnight.

BEFORE SERVING: Remove the napkin; place the dessert on a serving platter; decorate with walnuts and candied cherries. Keep refrigerated until you are ready for it. Serve cold.

SERVES 8.

· ·

Raspberry Cheese from Russia

The Russian cheese dessert, *pascha,* is primarily an Easter luncheon dish. But as it must be prepared well in advance, it makes a convenient and elegant party treat. When flavored with fruit preserves, it has an attractive color and a pleasant, refreshing taste.

2 pounds creamed cottage cheese
1 pound pure seedless raspberry preserves
½ pound sweet butter, soft
1 cup confectioners' sugar.
4 egg yolks
1 egg
1 pint sour cream

Press the cheese through a strainer. Mix with raspberry preserves, and press through a strainer once more.

Beat the butter with sugar for 5 minutes. Add egg yolks and the egg, one by one, still beating. Beat 5 more minutes.

Combine the cheese mixture with the creamed butter and sour cream. Mix very well. Refrigerate for 1 hour.

Now form mixture into a roll. Wrap it tightly in 2 cotton napkins which have been dipped in water and squeezed out. Place something heavy on the top to encourage draining off of any liquids. Refrigerate overnight.

BEFORE SERVING: Remove the napkins. Place the cheese loaf on a serving platter, decorate with raspberry preserves. Refrigerate until you are ready for it. Serve very cold.

SERVES 16.

· ·

Afternoon Coffee

PREFACE

Hungarians like to say that good coffee should be as black as the devil, as hot as hell, and as sweet as a kiss. Indeed Europeans take their coffee with milk or cream only for breakfast; otherwise it is served black and very strong, often just in demitasse cups.

A century ago ladies in Vienna, Lucerne, or Copenhagen liked to visit each other in the afternoons and chat over coffee and cakes. It was really not much different from our coffee klatches except that Europeans of that era were more formal in their attire and dressed up elegantly for the occasion, while our housewives drop in on each other just as they are.

Today so many women work outside their homes in Europe that afternoon visiting is limited to Sundays and holidays. Sunday dinner is usually served around 2 p.m., and by four the lady of the house is ready to see her friends. It is an extremely pleasant way of entertaining, and requires little preparation and expense. There is nothing more satisfying than a Sunday afternoon in good company with the house fragrant with freshly brewed coffee and homemade pastry.

European cakes are less sweet than American, but often heavier, and it is customary to serve very small pieces. In Central and Northern Europe, the favorite old coffee cakes are made with yeast. This requires a little work because with a few exceptions the dough must be kneaded by hand, but still it is nothing to scare you away. With refrigeration, we can always get fresh yeast, and the thermostat on our oven guarantees the proper rising and baking of our cakes. I have included in this chapter a few recipes for yeast-dough coffee cakes that are simple and always successful.

Most suitable for serving with afternoon coffee are pastries that are nourishing but neither too heavy, nor very sweet. You will find 19 such recipes in this chapter. Most of them can be also used for luncheons; fruit tarts are especially nice for this purpose.

I have always found pastries that freeze well a real treasure. Egg Bread, Almond Muffins, Sour Cream Cake, Poppy Seed Rolls, Sultans' Cake, Easter Baba, Kings' Cake, Walnut Cake—all can be prepared well ahead on a boring rainy day and stored for use on an appropriate occasion.

Some of the recipes I have selected for afternoon coffee, like Almond Muffins, Sour Cream Cake, Sultans' Cake, Kings' Cake, Blueberry Cake, and Almond Coffee Cake, are whipped up in a jiffy in one mixer bowl, and it takes no more effort or skill than when you are baking something from a mix. But the result is very different.

You have to be a very experienced and gifted cook to let

yourself forget the measuring cup and spoons. Our grandmothers seldom used them, and they could not regulate the
temperature of baking or roasting as well as we can. But
skilled cooks who really mastered the art of preparing food
were a rarity in those days and they were famous in their
communities just because there were so few of them. Today
every hostess can produce perfect pastry if she is willing to
follow recipes precisely.

· ·

Accompaniments for Afternoon Coffee

1. Coffee Vienna Style
2. Bulgarian Yogurt Drink
3. Jewish Egg Bread
4. French Almond Muffins
5. Sour-Cream Cake from Finland
6. Poppyseed Roll from Czechoslovakia
7. Sultans' Cake from the Balkans
8. Polish Easter Baba
9. Rum Baba from Prague
10. German Kings' Cake
11. Three Kings Cake from Paris
12. Jewish Passover Walnut Cake
13. Rumanian Charlotte
14. Prune Rolls from Moravia
15. Prune Tart from Vienna
16. Fruit Tart from Luhacovice
17. Blueberry Cake from Slovakia
18. Cherry Tarts from Saxony
19. Plum Tarts from Slovakia
20. Peach Tarts from Slovakia
21. Danish Almond Coffee Cake

· ·

Coffee Vienna Style

This rich coffee is more a dessert than a beverage. The Viennese drink it in small sips without mixing. As it slowly melts, the ice cream will mix itself with the coffee.

Serve with tiny cookies or small pieces of light pastry.

8 cups cold water
½ cup freeze-dried coffee
1 pint vanilla ice cream
1 cup whipping cream
2 teaspoons confectioners' sugar

Mix the water with the coffee in a jar. Refrigerate for several hours. Pour coffee into 8 tumblers. Divide ice cream into 8 portions, add to the tumblers. Top with cream whipped with confectioners' sugar. Serve immediately.

SERVES 8.

Bulgarian Yogurt Drink

For people who do not like coffee or for the younger set, it is nice on a hot summer day to have a Bulgarian variation of our milk shake.

Surprise your friends with this original yogurt drink.

Juice 2 oranges
Grated rind ½ orange
2 tablespoons honey
1 tablespoon confectioners' sugar
1 cup milk
1 pint plain yogurt

Refrigerate all ingredients for several hours. Whip everything in a blender. Serve in tumblers.

YIELDS 1 QUART.

Jewish Egg Bread

When I was a little girl in Warsaw, my mother often took me on Friday afternoons to a Jewish bakery where delicious egg bread called *chalka* was sold for the coming Sabbath. The shiny loaves still warm, just brought from the hearth, filled the street with a tempting aroma. Oh, how I used to love it! I hardly could wait to get home and eat those thick slices spread with sweet butter, accompanied by hot cocoa.

Making egg bread at home takes some effort, but the result is certainly worth it.

1 ounce yeast
⅔ cup sugar
2 ¼ cups tepid water
¾ teaspoon salt
8 cups flour, sifted
¼ cup salad oil
3 eggs, well beaten

Combine yeast with 1 tablespoon sugar, stir in ½ cup water. Place in a 100° oven for 15 minutes. Add the flour. Combine with remaining sugar mixed with salt. Add oil, 2 eggs, and enough water to make an elastic dough. Knead until smooth. Place in a greased bowl, cover with a towel. Let rise in a 150° oven until doubled in size. Take out, punch down. Divide into 2 parts. Divide each part into quarters. Form rolls out of each 3 quarters. Braid each set of 3 rolls beginning at the center and working toward each end. Press the ends tightly together. Divide each remaining quarter into 3 parts. Make rolls and braid each set into two strands. Press them into the top of the two loaves you have just made. Place the loaves on a greased cookie sheet. Cover with a damp towel and let rise in a 150° oven for ½ hour. Take out, spread with remaining egg. Bake in 350° oven for 45 minutes until golden brown.

YIELDS 2 LARGE LOAVES.

French Almond Muffins

When you look through the illustrated books of French gourmet cooking, you admire the artistic, elaborate decoration of particular dishes. They are all very beautiful and surely no less tasty, but these are productions of the sizable staffs of great French restaurants, or of famous chefs preparing banquets and feasts. These elaborate concoctions do not have much to do with the cookery of French working girls and the average French family, which is much, much simpler, but nevertheless quite elegant and savory.

Just try these little muffins from everyday French cuisine and see how suitable they are for your needs.

¼ pound sweet butter, soft
1 tablespoon sugar
4 eggs
1 teaspoon vanilla
½ cup flour
½ cup corn starch
1 teaspoon baking powder
3 ounces almonds, ground
½ cup confectioners' sugar

Beat butter with sugar until creamy, add eggs one at a time, and vanilla, still beating. Add flour combined with corn starch and baking powder. Beat for 5 minutes. Fold in almonds.

Fill well-buttered muffin pans ⅔ full. Bake at 350° for 25 minutes. Cool slightly, remove from pans. Sprinkle with sugar through sieve.

SERVES 8.

. . . .

Sour-Cream Cake from Finland

On Christmas day in Finland, afternoon coffee is often served buffet style. On a table covered with a red linen cloth, a blooming cyclamen and candles are placed at one side, cups and saucers at the other. A sour-cream cake stands in the center. Both ends of the table are also used for platters with assorted cookies and for family heirlooms—a shiny copper coffee kettle and pitchers and jars with sugar and cream.

1 pint sour cream
1 teaspoon vanilla extract
1 teaspoon baking soda
1 cup potato flour or corn starch
1 ½ cups sugar
1 ½ cups flour

Whip the cream for 5 minutes, add other ingredients in given order, beating. Beat 5 more minutes. Pour into a tube pan that has been buttered and sprinkled with bread crumbs. Bake in a 350° oven for 1 hour.

SERVES 10.

- -

Poppyseed Roll from Czechoslovakia

During my childhood in Poland, I thought in my naïveté that poppyseed rolls, one of our traditional Christmas pastries, were a purely Polish specialty. Later with great surprise I found similar delicacies in many Central European countries. This is how poppyseed rolls are prepared in Czechoslovakia.

Do not be tempted to roll out your dough too thin. If you do, the roll will break on the top during baking.

DOUGH
⅓ cup butter
2 cups and 2 tablespoons sifted flour
½ cup confectioners' sugar
1 egg
2 egg yolks
¼ cup coffee cream
½ teaspoon vanilla extract
½ tablespoon grated lemon rind
1 ounce yeast
½ tablespoon sugar

FILLING

1 ½ pounds poppyseed pastry filling
1 tablespoon candied orange rind, finely chopped
1 tablespoon almonds, finely chopped
½ tablespoon grated lemon rind
2 tablespoons raisins

ICING

½ cup confectioners' sugar
1 tablespoon rum

TO MAKE DOUGH: Cut the butter into the flour with a knife, and rub in with fingertips. Combine with confectioners' sugar. Add all the other ingredients and yeast mixed with sugar. Knead the dough until smooth. Roll out a rectangle ½ inch thick.

Combine the ingredients of the filling. Spread over the dough. Roll.

Place seam down in a well-buttered narrow loaf pan or tube pan. Cover with a dish towel. Let stand in 150° oven till the roll doubles in size—about 1½ hours. Bake uncovered in 350° oven for 45 minutes. Cool. Remove from the pan.

TO ICE: combine the ingredients and spread over the roll.

YIELDS 1 ROLL.

SERVES ABOUT 10.

Sultans' Cake from the Balkans

Surprisingly, my Sultans' cake does not come from Turkey. It is a popular Sunday treat in Balkan nations, whose people learned to love the Turkish delicacies brought them by Turkish merchants. Visitors from the distant land of harems and lavish feasts were often invited to sip coffee with the family around the kitchen table and to tell of the fairy tale riches of Turkish sultans and the beauty of their wives and daughters. Cocoa, almonds, raisins, and the exotic spices the Turkish visitors left were used for baking rich cakes worthy of a sultan's table.

BATTER
½ pound margarine
2⅔ cups confectioners' sugar
1 teaspoon vanilla extract
4 eggs
4 cups flour, sifted
4 teaspoons baking powder
4 tablespoons cocoa
1 cup milk
2 tablespoons orange brandy or rum
6 ounces raisins
4 ounces walnuts, chopped

SPREAD
12 ounces chocolate chips
3 tablespoons butter
2 tablespoons milk

TO MAKE BATTER: Beat the margarine adding sugar in small parts. Add vanilla. Beat 5 minutes. Add the eggs one at a time, beating at high speed. Beat for 5 more minutes.

Mix the flour with baking powder and cocoa. Add to the batter in small portions, alternately with milk, beating constantly. Add brandy and beat for 5 more minutes.

Mix the raisins with the nuts and fold into the batter. Bake at 350° for 1 hour and 15 minutes in a tall 9-inch ring pan that has been buttered and sprinkled with bread crumbs. Cool.

TO MAKE SPREAD: Melt the chocolate with butter and milk over very low heat. Do not let it boil. Pour over the cake and let it drip over the sides.

This cake is even better the following day.

YIELDS 1 TALL 9-INCH CAKE.

SERVES 15.

. . . .

Polish Easter Baba

Polish Easter is a very fragrant holiday. The whole house is fragrant with freshly waxed floors. Hyacinths stand on a white tablecloth next to various cakes, sweetly smelling of vanilla, almonds, rum, chocolate, and candied fruits. The queen among these cakes is a tall *baba* in a snowy bonnet of icing.

This recipe never fails, and is a real labor saver. It is a yeast cake that requires no kneading, because it is done in a mixer. This cake freezes very well.

Prepare all your ingredients before you start mixing. Do not rely on pre-sifted flour. Sift it yourself for exact measurement.

2 small cakes yeast
1 cup milk, lukewarm
4 cups flour, sifted
½ pound margarine
¾ cup sugar
5 eggs
1 teaspoon almond extract
Rind 1 lemon, grated
1 cup raisins
⅓ cup almonds, chopped in a blender

ICING
2 cups confectioners' sugar
4 tablespoons rum

Dissolve yeast in milk, mix with 1 cup flour. Place in the oven, heated to 150°, to rise until doubled.

Cream margarine; gradually beat in sugar. Add eggs one at a time, still beating. Beat for 5 minutes. Add yeast mixture and 2 cups flour. Beat for 10 minutes. Add the remaining flour, almond extract, and lemon rind. Beat 5 more minutes. Fold in raisins and almonds. Place in a 9-inch round, 4-inch tall tube pan that has been buttered and sprinkled with bread crumbs. Cover with a dish towel and let rise in 150° oven for 1 hour. Take out, quickly heat the oven to 350°. Bake uncovered for 45 minutes. Cool. Remove from the pan.

TO MAKE ICING: Combine the sugar with rum. Spread over the baba. Cut in generous slices.

SERVES 15.

. . . .

Rum Baba from Prague

I spent the last winter of World War II in Prague. It was a difficult time with food and fuel shortage, frightening air raids, and a nagging worry for the safety of my loved ones left in Poland. But I'll always remember that winter with nostalgia because of the warmth of the friendly hearts of my Czech friends.

After a miserable couple of months in rented rooms, my sister and I landed a painter's tiny studio apartment with a window in the ceiling. Here is what a dear Czech friend brought us for housewarming.

BATTER

⅓ cup margarine, melted
¾ cup sugar
2 eggs
1½ cups flour
2 teaspoons baking powder
3 tablespoons milk
1 grated orange rind

SYRUP

1 cup sugar
½ cup water
¼ cup orange juice
¼ cup rum

TO MAKE BATTER: Place all the ingredients in a bowl, beat 5 minutes at medium speed. Bake in a well-buttered 8-inch fluted ring pan in 350° oven for 45 minutes.

TO MAKE SYRUP: Cook the sugar and water to a heavy syrup. Add orange juice and rum.

Remove the warm cake from the pan onto a serving platter and immediately pour the syrup slowly over it. Cool thoroughly before cutting.

SERVES 10.

. . . .

German Kings' Cake

This cake is traditionally served in Germany at the very end of Christmas season celebrations, which last until January 6, the Feast of the Three Kings or Twelfth Night. That is also the date on which Christmas decorations are put away and the tree is thrown out. It could be a sad day for the children,

if it were not for special treats and possibly an afternoon party
with games.

The lovely Christmas tree custom originated in Germany
a century ago, and was soon adopted by the whole Christian
world. The same feeling of melancholy lingers everywhere at
the sight of a Christmas tree lying desolately in the backyard
with most of its needles gone, a few forgotten silver threads
hanging here and there as reminiscence of its by-gone glory.
But the delicious aroma of a home-baked cake helps to make
up for the loss of the tree.

½ pound margarine, soft
1 ½ cups confectioners' sugar
4 eggs
1 teaspoon vanilla extract
4 tablespoons rum
2 ¾ cups presifted flour
1 cup cornstarch
3 teaspoons baking powder
⅓ cup milk
6 ounces raisins
3 ounces almonds, coarsely chopped
2 ounces candied orange rind, finely chopped

Beat the margarine with sugar for 5 minutes. Add the
eggs one at a time, beating at high speed. Add vanilla and rum,
beat some more. Add the flour, cornstarch, baking powder,
and milk; beat 5 minutes more. Fold in raisins, almonds, and
orange rind.

Bake at 350° for 1 hour and 15 minutes in a 9-inch round
deep tube pan that has been buttered and sprinkled with bread
crumbs.

Sprinkle the hot cake with confectioners' sugar through
a sieve. Cool. Keep covered. Cut into generous slices with a
sawing motion. It tastes better the day after baking.

SERVES 15.

Three Kings' Cake from Paris

In many European countries Three Kings Day (Epiphany) on the 6th of January is an official holiday and an occasion for small gatherings of family and friends. In Paris on that day it is customary to serve a delicious cake of puff pastry stuffed with almond paste.

We can make such a cake with very little effort, using strudel dough, and a ready-made almond paste. Both are sold in many larger supermarkets.

To add to the holiday fun, place one whole almond in the paste as the French do. Whoever finds it in his piece of cake is crowned the King or the Queen of the day and receives a small gift.

This cake is best when eaten the same day it is baked.

1 can almond paste
1 ½ tablespoons rum
4 sheets strudel dough
¼ cup butter, melted
1 whole almond, peeled
1 egg
1 teaspoon water

Mix almond paste with rum.

Place 1 sheet of the dough on a buttered 12-inch round pizza pan. Fold the hanging sides inside to make a good fit. Sprinkle with half of the butter. Cover with second sheet and fold to fit again. Spread with almond paste. Place the almond in the paste. Cover with the third sheet; sprinkle with remaining butter. Cover with the last sheet; fold over the sides. Glaze the cake with the egg mixed with water. Press the sides together.

Make a small vent in the center. Place in a 400° oven for 20 minutes. Reduce the heat to 375° and continue baking for another 25 minutes until the pastry is brown and crisp.

YIELDS ONE 12-INCH ROUND CAKE.

SERVES 8.

. . . .

Jewish Passover Walnut Cake

Passover and Easter coincidentally fall close to one another. In our ecumenical times, it might be interesting to try during our own holiday the traditional delicacies of a neighbor of a different faith.

8 eggs
¼ teaspoon salt
½ teaspoon cream of tartar
1 ½ cups sugar
Peel grated from 1 orange
⅓ cup orange juice
½ cup potato or corn starch
½ cup matzo cake meal
1 cup walnuts, toasted, ground
½ cup confectioners' sugar

Beat 7 egg whites with salt and cream of tartar until stiff but not dry. Add half of the sugar by spoonfuls, still beating. Beat for 2 more minutes.

Beat 7 egg yolks with the remaining whole egg and sugar for 5 minutes. Stir in orange peel and juice. Sift matzo meal with starch. Add ¼ of the egg whites to the egg yolk mixture, mix lightly. Pour this mixture over the remaining egg whites.

Sift the matzo meal mixture over it. Mix lightly. Fold in the nuts.

Bake in a 10-inch ungreased tube pan in 350° oven for 1 hour. Cool in the pan. Remove. Sift the sugar over the cake.

YIELDS ONE 10-INCH CAKE.

SERVES 10.

. . . .

Rumanian Charlotte

The Rumanian people are descendants of an ancient tribe conquered by the Romans who exiled their convicts to Rumanian territory. Today's Rumanians are a mixture of all the Central European nations. They are a peaceful people, mostly farmers. The Rumanian intelligentsia has been strongly influenced by French culture for centuries. Many popular Rumanian dishes carry French names, just like this tasty apple tart.

DOUGH
½ pound margarine
4 cups sifted flour
1 cup confectioners' sugar
4 teaspoons baking powder
2 eggs
1 egg yolk
⅓ cup sour cream
Grated rind of one lemon
1 teaspoon vanilla extract

FILLING
4 cups tart apples, peeled, shredded
⅓ cup sugar
1 tablespoon cinnamon

ICING
2 cups confectioners' sugar
4 tablespoons lemon juice

TO MAKE DOUGH: Cut the margarine into the flour with a knife and rub in with fingertips, add the sugar and the baking powder. Mix. Add the remaining ingredients and knead the dough. Refrigerate for ½ hour.

Roll out two 9″ × 12″ rectangles. Place one in a pan that has been buttered and sprinkled with bread crumbs. Bake in 375° oven for 15 minutes.

TO MAKE FILLING: Combine all ingredients; spread over the cake. Cover with second rectangle. Return to the oven for 45 minutes. Cool. Remove from the pan.

TO MAKE ICING: Combine the sugar with lemon juice; spread over the cake. Refrigerate. Cut into 32 squares.

YIELDS 32 SQUARES.

· ·

Prune Rolls from Moravia

Czechs love pastry, particularly when it is made of yeast dough and stuffed with a fruit or cheese filling. Baking with yeast takes a little more time, but the result is worth the effort. Women in Moravia make these delicacies quite often, sometimes two or three times a week. With our busy lives, we would rather save this treat for special occasions.

DOUGH
1 package dry yeast
2 tablespoons warm water
4 cups sifted flour
¼ cup sugar
½ teaspoon salt
1 teaspoon grated lemon rind
¾ cup butter, soft
3 egg yolks
1 cup coffee cream

FILLING
1 jar prune pastry filling
2 tablespoons candied orange rind, finely chopped
½ cup walnuts, finely chopped
¼ teaspoon cinnamon

½ cup confectioners' sugar

Dissolve yeast in warm water. Sift the flour with sugar and salt. Add lemon rind.

Cream the butter; add egg yolks; beat some more. Add cream. Add yeast and butter mixture to the flour. Knead the dough until smooth. Refrigerate in a covered bowl overnight.

The next day roll dough out ⅛ inch thick on a lightly floured board. Cut into 3-inch squares. Place 1 teaspoon of filling in the center of each square. Fold each one envelope fashion and pinch to seal the edges.

Place on a buttered baking sheet seams down, cover with a light dish towel. Place in 150° oven until doubled in size. Bake in 375° oven for about 15 minutes. Remove to a rack. Sprinkle with confectioners' sugar through a sieve.

YIELDS 3½ DOZEN.

Prune Tart from Vienna

This cake, elaborate in appearance and rich in flavor, is very simple to make. You can cook your own prune filling or use a prepared one to save labor. Almond paste is available in most larger food markets.

FILLING
3 cups pitted prunes, finely chopped
¾ cup sugar
½ cup water
½ teaspoon grated lemon rind

DOUGH
¾ pound margarine, soft
1 cup sugar
5 egg yolks
1 cup almonds, ground
4 cups sifted flour
½ cup almond paste

TO MAKE FILLING: Cook all the ingredients at low heat, stirring until the filling becomes very thick. Cool.

TO MAKE DOUGH: Cream the margarine with sugar until fluffy. Add egg yolks and beat 5 minutes more. Fold in almonds. Knead the dough, adding flour one cup at a time.

Combine ⅓ of the dough with almond paste and refrigerate.

Roll out the remaining dough on a buttered 12″ × 15″ cookie sheet, forming a 11″ × 13″ rectangle. Spread with the filling. Crumble the chilled dough over it evenly. Bake in 325° oven for 50 minutes until golden brown.

SERVES 12.

. . . .

Fruit Tart from Luhacovice

It was a hot July, the last summer of World War II, a restless time at a Czech spa in the Carpathian Mountains. Often in the dark, guerrillas would descend through the wooden slopes to help themselves to whatever they needed from the small homes perched in the valley.

A sudden noise woke me up in the middle of the night. Someone had jumped from the porch through my open window. Too frightened to move, I watched the dark figure leaning over the table where I had left the purse with all my money and my sapphire ring. After looking around for a moment, the intruder grabbed the fruit tart which my landlady had baked for us earlier that evening, and before I dared to breathe—he was gone.

BATTER
1 package dry yeast
½ cup milk, tepid
⅓ cup sugar
2 cups flour
¾ stick sweet butter
2 eggs, separated
1 teaspoon vanilla
Pinch salt

TOPPING
4 cups sliced fruit (peaches, apricots, prune plums, or
*   apples)*
¾ cup confectioners' sugar

Combine the yeast with milk and 1 tablespoon sugar. Let stand for 5 minutes. Add ½ cup flour, and place in 150° oven until it doubles in size.

Cream the butter with the remaining sugar, add the egg yolks and vanilla, and beat for 5 minutes. Combine with the yeast. Add remaining flour in small portions, beating. Beat 5 more minutes. Fold in egg whites whipped with salt.

Pour into a 9″ × 12″ baking pan that has been buttered and sprinkled with bread crumbs. Cover with a light towel, and place in 150° oven until batter doubles in size. Cover with sliced fruit and bake in 350° oven for 30 minutes. Cool. Cut while still in the pan. Sprinkle with the sugar through a sieve before serving.

SERVES 10.

. . . .

Blueberry Cake from Slovakia

On both sides of the Carpathian Mountains, millions of blueberry bushes grow in abundance, to the delight of the natives. For centuries the country children have spent their summer days gathering blueberries for the family table or to sell at farmers' markets. Their berries are smaller than those we see in American supermarkets; the taste is somehow more intense and the juice bluer. But American blueberries serve as well for this delicious coffee cake which I learned to make one summer in Slovakia.

½ pound margarine
1 ¼ cups sugar
4 eggs
2 cups flour
1 ½ teaspoons vanilla extract
2 teaspoons baking powder
1 ½ pints blueberries
½ cup confectioners' sugar

Beat the margarine with the sugar for 5 minutes. Add the eggs alternately with the flour; beat 5 more minutes. Add vanilla and baking powder; beat 1 minute more.

Fold batter into a 9″ × 12″ pan, buttered and sprinkled with flour. Distribute blueberries evenly over it. Bake in a 400° oven for 45 minutes.

Remove from the oven. Cool. Sprinkle with confectioners' sugar through a sieve. Cut in the pan and transfer to an attractive platter.

SERVES 15.

· ·

Cherry Tarts from Saxony

Each Saturday German housewives like to bake something sweet to offer relatives and friends who come calling on Sunday afternoons. These Sunday gatherings with coffee and home baked goodies are a true national tradition in Saxony.

This is a typical German summer treat made with sour or sweet black cherries, blueberries, or sometimes rhubarb.

DOUGH
3 cups flour, sifted
4 teaspoons baking powder
⅔ cup sugar
¾ cup margarine, soft
1 teaspoon vanilla
2 eggs
3 tablespoons milk

TOPPING

½ cup bread crumbs
2 pounds cherries, pitted
1 cup flour, sifted
¼ pound butter, soft
1 cup confectioners' sugar
½ teaspoon cinnamon

½ cup confectioners' sugar

TO MAKE DOUGH: Combine the flour with baking powder and sugar. Cut margarine into the flour, add remaining ingredients, and knead the dough for a few minutes. Spread the dough in a 9″ × 12″ pan that has been buttered and sprinkled with bread crumbs.

TO MAKE TOPPING: Sprinkle the dough with bread crumbs. Scatter cherries evenly over it.

Knead the flour with butter, sugar, and cinnamon. Crumble it and sprinkle on the fruit. Bake in 350° oven for 50 minutes.

Take out and sprinkle with confectioners' sugar through a sieve. Cool. Cut into square tarts while still in the pan.

SERVES 12.

Plum Tarts from Slovakia

Prune plums are the favorite fruit of Slovakia. They are used to make a delicious vodka, for soups and compotes, and of course as fillings and toppings for dumplings and pastries.

Choose medium-sized prune plums, sweet and ripe, but still firm. The tarts taste best when eaten the same day. Serve with whipped cream in a separate bowl.

DOUGH
1 ½ sticks margarine
2 ¾ cups sifted flour
⅔ cup confectioners' sugar
2 teaspoons baking powder
1 egg
1 egg yolk
3 tablespoons sour cream

TOPPING
25 prune plums, cut lengthwise into halves

ICING
1 cup confectioners' sugar
2 tablespoons rum

TO MAKE DOUGH: Cut the margarine into the flour and rub in with fingertips. Add the sugar and baking powder. Combine with the egg, egg yolk, and sour cream. Knead the dough for a few minutes.

Divide the dough into 4 parts. Place on a buttered 12″ × 15″ cookie sheet. Spread the dough till the sheet is almost covered. Arrange the prunes in rows, skin down. Bake in 375° oven for 45 minutes. Cool.

TO MAKE ICING: 1 hour before serving, mix sugar with rum. Spread over the cake. Cut into 25 tarts.

YIELDS 25 TARTS.

· ·

Peach Tarts from Slovakia

Peach tarts are a real summer delight. Choose ripe but firm peaches that can be peeled easily. These tarts are best when fresh. Bake a few hours before your guests arrive. Sprinkle with sugar just before serving.

DOUGH
*Prepare the dough as for Plum Tarts**

TOPPING
8 large freestone peaches
½ cup confectioners' sugar

Skin and quarter the peaches. Place on the dough in rows. Bake in 375° oven for 45 minutes. Cool.

Cut into 32 tarts. Sprinkle with sugar through a sieve just before serving.

YIELDS 32 TARTS.

. . . .

Danish Almond Coffee Cake

Scandinavian cuisine is famous for its pastries—not without cause. The Scandinavians are sociable and hospitable people. No matter what time of the day you happen to visit a Danish or Swedish home, you may be sure that a traditional cup of strong coffee with a freshly baked piece of something sweet and delicious will be served.

The following recipe makes it possible to serve an attractive coffee cake in one hour from the time you enter the kitchen.

BATTER
4 eggs
2 cups sugar
2 cups flour
3 teaspoons baking powder
¼ pound margarine
1 cup milk

TOPPING
1 cup slivered almonds
¼ pound butter
½ cup sugar
Dash of salt
1 tablespoon instantized flour
½ cup coffee cream

TO MAKE BATTER: Beat the eggs with the sugar for 5 minutes. Add the flour and baking powder. Melt the margarine in hot milk, pour into the batter. Beat for 5 minutes. Bake in 350° oven for 35 minutes in a 9″ × 12″ pan, buttered and sprinkled with bread crumbs. Cool slightly.

TO MAKE TOPPING: Put all the ingredients in a saucepan. Simmer for a few minutes, mixing. Spread evenly over the cake. Place the cake under broiler for 5 minutes or until golden. Cool. Cut in the pan.

YIELDS ONE 9″ × 12″ COFFEE CAKE.

SERVES 10.

Dinner Parties

PREFACE

If you ever intend to see Europe, do not go on an organized tour. Nine capitals in seven days isn't very funny. After the eighth cathedral, the poor traveler cares about nothing but a hot bath.

It might be better to get acquainted with the culinary customs of different nations by using recipes from the following chapter for your dinner parties. These traditional, original dishes will take you on a pleasant journey without the necessity of buying a plane ticket.

But if this culinary experimentation makes you wish to see the faraway places you have heard so much about, you will not get any real feeling of the country you visit if you travel in a large group of fellow Americans and never mingle with the natives. When you go to a new country, meet the people, laugh with them, sing with them, and eat with them. Language is no longer a barrier, since many Europeans speak English, and many young Americans know at least a little of some foreign tongue.

My old professor at the Sorbonne used to say that one does not absorb anything after spending longer than an hour at a time in a museum. It is better to stroll in the city and get lost, or to sit in a sidewalk café watching the street vibrate with life than to see all the famous castles through conveniently tinted bus windows.

If you are interested in native food, you will do better to stay at small inns and eat in small, local restaurants than in large international hotels that look the same everywhere, and where you will meet only other tourists like yourself.

Once in Warsaw I observed a group of Canadian tourists shifted to a separate room of a well-known restaurant. There, behind the closed doors, all together just if they had never left Toronto, they waited endlessly for their inferior meal, even without the opportunity of observing how local people dress, eat, and enjoy themselves. Service was poor because waiters did not expect individual tips. The group was not shown the daily menu, but was served boiled beef with boiled potatoes, which very few liked. They complained about how bad Polish restaurants and Polish food were.

At the same time in the main room of this restaurant, for a reasonable price, I was given a choice of hare in sour cream, goose and cabbage, duck with apples, and famous Polish beef rolls in wild mushroom sauce. All this was followed by a selection of seven different tortes. The service was prompt and the meal delicious.

I heard once of a group on an organized, budget-priced excursion that was served similar hamburgers with catsup in

Vienna, Rome, and Paris because the travel agent had said that it is what Americans like best.

Europe offers many charming little places to sleep and eat, but if you can, avoid traveling in the middle of summer with all the crowd. You will have to wait constantly for everything, and you will get a decent meal nowhere. Your opinion on continental cuisine may thus become quite distorted.

As soon as you arrive, inquire in your hotel where local people like to eat, go there and order local dishes. Don't ask for eggs and bacon or American cereal for breakfast in Vienna. Forget the steak, try native dishes in Budapest. Start your Swedish smorgasbord with herring. Ask for seafood in Spain, not in Czechoslovakia. Wherever you eat, inquire for the specialty of the house and the dish of the day. It will be fresher and better than others on the menu, and probably not too expensive.

Investigating small taverns in Marseilles, letting a little *bambino* guide take you to his uncle's favorite inn in Sicily, dropping in a café which looks enchanting even if no one ever heard of it—can be a stimulating and rewarding experience. Do not forget to take along a pocket notebook in which to record names of dishes and wines you have enjoyed and maybe one or two recipes you may be lucky enough to get.

After testing the continental cuisine in its original setting, you will want to recreate your favorite discoveries, and soon you may become a real connoisseur of European dishes. This book will be your culinary guide to delightful dinner parties. Each new recipe you try, each new adventure in dining will help you to expand and enrich your taste and your guests' taste as well.

Like Americans, Europeans love to talk sitting around the table. Conversation will never be better, more sparkling than when there are just 6 to 10 persons present. A dinner party is at its best in small intimate groups with flickering candlelight, flowers on the table, and soft music in the background. But most important of all is a relaxed hostess, confident in her art of cooking and serving.

Planning a dinner party, the hostess must choose dishes that can be cooked ahead of time; only one or two items should be left for the last moment. For instance, when meat just has to be reheated, you can easily serve freshly cooked vegetables and tossed greens. In case the main dish requires special attention, it is sensible to prepare a molded salad, potato salad, or any other convenient kind that can be made the previous day or in the morning. One of my friends, an experienced hostess who gives wonderful dinner parties, makes the dessert and freezes it at least a week ahead. She also has a habit of baking tortes whenever she has some free time, and then she has them ready and waiting in the freezer when she starts planning the menu for her next dinner party.

Give your family simple meals on that day, set the table for your dinner party early, if possible at noon. Do not do any last moment cleaning, but treat yourself to an hour's afternoon rest. A nap does wonders for the good humor and glamorous appearance of the hostess.

Music is helpful in livening up social gatherings, but it should never dominate the conversation. Choose your records before the guests arrive. Pleasant melodies without singing are best, unless you are planning a concert.

Sometimes an eager hostess trying to combine everything at once serves several cocktails and plenty of snacks when a sit-down dinner is planned. Real connoisseurs never do this, as they believe that it would spoil the appetite and the ability of their guests to appreciate the food.

In Europe hard liquor is seldom served before dinner. Apéritif wines, such as vermouth or sherry, which have a little higher alcoholic content than table wines, are often served instead. If your friends like a drink, offer it, but serve a second one only if someone requests it. Do not have a regular cocktail party before you call your guests to the table.

Restrain yourself from constantly putting things in place. A full ashtray and some forgotten glasses here and there hurt less than an overly eager hostess. If you can go to another room after lunch or dinner, do not even bother to clear the table, just put away the food. Never do dishes or even fill the dishwasher while your guests are still there, and never let your friends help you clean up. They did not put on their good dresses in order to mess them up in the kitchen. They came to relax and enjoy themselves—not to work.

THE ART OF SERVING WINE

In sunny Italy, France, and Spain grapes grow sweet, fragrant, and healthy. Wine is part of every man's life, his joy, his pleasure. One cannot imagine a meal in southern Europe without this nectar of the gods. Bacchus crowned with grape leaves, born of fire and nursed by rain, the king of wine, was god of happiness and merry times.

Some young, inexpensive brands of wine used in those countries for everyday may not be great, although genuinely pleasant. But for special occasions the selection is made with the greatest of care. No money or effort is spared for the right choice of wine and for its proper handling.

In the northern and eastern parts of Europe where there are no good vineyards, imported wine is a special treat. Tall glasses glowing with burgundy or a golden shimmering liquid adorn every elegant dinner table. Wine enhances the flavor of food and makes each meal festive. The small percentage of alcohol one or two glasses contain cannot intoxicate anyone easily, but is enough to stimulate conversation and produce good humor. Wine is beneficial for the digestion, for relaxation, and in many cases even for the heart.

There are no rigid rules to tell you what kind of wine must be served with what food. The two should blend in perfect harmony. The right choice comes with experience. Of course each person may have different preferences, but most beginners can use a few hints on selecting wines.

Red wine is made from red grapes fermented with their skins. When the juice is drained and fermented alone, a rosé wine results, with a lighter flavor. White wines are obtained from green grapes or sometimes red ones that did not ripen properly.

Wines with a little higher alcoholic content, such as sherry, vermouth, or port, are served before meals. If you wish, you may have champagne with your main dish, but it is primarily a drink for toasts and celebrations. Weddings, anniversaries, and New Year's Eve parties are always associated with the pale golden, bubbly beauty and delicate taste of champagne. When a meal will follow, a dry (not sweet) or semi-dry variety is best. Sweeter kinds are more appropriate after dinner.

Table wines should not be sweet. In fact, the red ones seldom are, but when choosing white wine, one must be careful because some, like Chablis, may be extremely dry, while others have varied degrees of sweetness.

Light white wines go well with fish and other seafoods. Chicken, turkey, and veal dishes call for white or light rosé. Red table wines are best with steak, roast beef, venison, and other dark meats. When you serve assorted cheeses with crackers, dry or semi-dry wines will go well. Whenever you are in doubt choose rosé. Sweet wines are left to accompany desserts and fruits. Bordeaux Sauternes are popular among the sweeter French wines.

The quality of wine is determined by the vineyard, its soil, the type of vine, and the skills of the producer. Quite a lot depends also on the weather. The crops of some years are recognized as great vintages. Those wines mature slowly, taking many years before they are at their best.

It is impossible to know all the factors in the making of a great vintage. Only top specialists are sufficiently competent to understand these subtleties. In our selection, therefore, we have to depend on the merchants.

The fame of French and Italian wines is justified. Many people also favor the excellent light German Rhine wines. Tokay is a unique, delicious Hungarian specialty. Comparatively inexpensive Portuguese rosé can nicely complement a most elegant dinner. Some of our American wines have a very pleasant bouquet and refined, delicate taste, although different from the French ones, even if they carry the same name. Good domestic wines are never sold at a very low price.

It would be wasteful to use expensive wines in cooking, but any wine that is not tasty enough for drinking is not good for cooking either. Cooking wine is always dry; it may be red or white, depending on the recipe. Wine does not add much to the caloric count of a dish as ½ cup contains only approximately 70 calories, less than most fruit juices. Wine sauce should be allowed to boil for a few minutes until all the alcohol evaporates and its odor vanishes. This is extremely important for obtaining the right flavor.

Most red wines are best when they are not chilled. However, when the directions on the label tell you to serve the wine "at room temperature," this does not mean at the temperature of an average American home, but rather a little below 60°. Europeans like their white wines moderately cold.

Good wine should never be removed from its original bottle before serving. It tastes best when the bottle has been opened an hour before dinner, and large, tulip-shaped glasses are only half-filled, because the full strength of the aroma develops in contact with air. It is not proper to serve water together with wine unless someone requests it.

Appreciation of fine wine requires a little ceremony in the way it is served; this elevates the mood of the occasion. The host pours the wine when all the guests have been seated. A small napkin wrapped around the neck of the bottle prevents dripping. The host first pours some into his own glass to catch any particles of the cork that may have fallen inside the bottle. Then he sniffs his wine and tastes it to evaluate its aroma and quality, and only after he finds it satisfactory does he fill the glasses of his guests, starting with the lady at his right.

Connoisseurs admire the beautiful color of wine and swirl it in their glasses to intensify the oxygenation before they drink it. They sniff it delicately with an expression of delight. Europeans believe that it is courteous to discuss the quality, the texture, and the bouquet of the wine they drink, and to compliment the host on his choice.

. . . .

For Dinner Menus

SOUPS AND SOUP GARNISHES
1. Hungarian Beef Bouillon
2. Slavic Red Beet Soup
3. French Provincial Onion Soup
4. Cabbage Soup from Finland
5. Hungarian Sauerkraut Soup
6. Cream of Carrot Soup from Finland
7. Scandinavian Summer Soup
8. Hungarian Peasants' Bread Soup
9. Spinach Soup from Oslo
10. Lemon Soup from Warsaw
11. Goulash Soup from Hungarian Countryside
12. Polish Blueberry Soup
13. Polish Plum Soup
14. Polish Strawberry Soup
15. Scandinavian Dried Fruit Soup
16. Jewish Matzo Dumplings
17. French Croutons

FISH
18. Stuffed Pike from Finland
19. Swedish Fish in Horseradish
20. Hungarian Fish Dinner
21. Baked Fish from Lake Balaton
22. Carp Jewish Style
23. Braised Fish Fillets from Norway
24. Jewish-Style Fish in Cream Sauce
25. Danish Fish with Cheese

MEATS AND GARNISHES
26. Roast Chicken Polish Style

27. *Chicken in Lemon Sauce from Warsaw*
28. *Turkish Chicken with Rice*
29. *Jewish Chicken with Grapes*
30. *Spanish Chicken with Vegetables*
31. *Hungarian Chicken with Paprika*
32. *Roast Pheasant or Grouse the Basque Way*
33. *Jewish Liver with Oranges*
34. *Roast Veal the Russian Way*
35. *Veal Dinner from Zürich*
36. *Veal Schnitzel from Vienna*
37. *Hungarian Veal and Paprika*
38. *Pork Chops from Poland*
39. *Roast Pork Loin from Moravia*
40. *Lamb in Cabbage from Norway*
41. *Leg of Lamb from Spain*
42. *Hungarian Wooden Plate Dinner*
43. *Hamburgers from Prague*
44. *Hungarian Beef Dinner*
45. *Norwegian Beef Stew*
46. *Roast Beef from Prague*
47. *German Thrifty Beef Roast*
48. *Spaghetti from Bologna*
49. *Hungarian Butter Dumplings*
50. *Hungarian Quick Dumplings*
51. *Swedish Mushroom Sauce*
52. *Sauce Hollandaise from Everywhere*
53. *A French Masterpiece—Sauce Béarnaise*
54. *Sauce Bordelaise from France*

VEGETABLES
55. *String Beans Polonaise*
56. *Carrots with Peas from Vienna*
57. *Spinach Creamed the Continental Way*
58. *Spinach the Belgian Way*
59. *Spinach Soufflé from Grenoble*

60. *Purée of Green Peas from Vienna*
61. *Finnish Christmas Rutabaga*
62. *Hungarian Baked Asparagus*
63. *Russian Red Beets*
64. *Hungarian Cabbage*
65. *Red Cabbage from Germany*
66. *Ukrainian Sauerkraut with Mushrooms*
67. *Noodles with Cabbage from Hungarian Countryside*
68. *Baked Potatoes from Finland*
69. *Swedish Spring Potatoes*
70. *Bohemian Boiled Potatoes*

PASTRIES
71. *Swedish Apple Squares*
72. *Hungarian Apple Strudel*
73. *Viennese Apple Strudel*
74. *Almond Strudel from Salzburg*
75. *Cheese Strudel from Vienna*
76. *Hungarian Walnut Strudel*
77. *Hungarian Poppyseed Strudel*
78. *Hungarian Blueberry Strudel*
79. *Cherry Strudel from Vienna*
80. *French Cherry Pie*
81. *Apple Pie from Switzerland*
82. *Strawberry Pie from Switzerland*
83. *Dutch Strawberry Cake*
84. *Swiss Chocolate Cake*
85. *French Easter Cake*
86. *Russian Cheese Cake*
87. *Italian Cheese Torte*
88. *Viennese Sacher Torte*
89. *Coffee Torte from Paris*
90. *French Torte Curaçao*
91. *Grape Torte from Moravia*
92. *Rum Torte from Vienna*

Hungarian Beef Bouillon

It seems that all around the world the ways of cooking bouillon are similar, but still wherever you go, it will taste a little different. The distinction is made by the different vegetables and spices used by various nations.

Once you have a good beef stock, you can easily cook many different delicious soups, using it as a base. Cook your bouillon a day ahead, refrigerate overnight, and when the crust of fat has formed, remove it before reheating your soup.

Serve beef bouillon with small noodles or dumplings that have been cooked separately. Let's see how the Hungarians make it!

2 pounds beef shank
1 pound beef bones
2 of each: carrots, parsley roots or parsnips, celery
    stalks, onions, and turnips
½ green pepper
1 clove garlic
4 peppercorns
1 bay leaf
2 quarts water
Salt to taste

Place all the ingredients in a kettle. Bring to a boil, skim, and simmer for 2½ hours. Strain.

SERVES 6.

Slavic Red Beet Soup

Soup made with red beets is an uncontested favorite with all Slavic nations. It is called *borsch* in Russia and *barszcz* in Poland and by various names elsewhere. It may be made with or without vegetables, with cream or clear.

Before lemons or vinegar were available in Eastern Europe, a mixture of flour and water was put aside to sour, to be used later in the soup. This is no longer done. We also no longer have to go to all the trouble of peeling and cooking beets to get excellent results. It takes only 15 minutes to prepare this soup.

6 beef bouillon cubes
3 ½ cups water
2 1-pound cans red beets
1 14-ounce can brown baked beans
1 tablespoon lemon juice
½ teaspoon salt
1 teaspoon sugar
3 tablespoons sour cream
1 tablespoon green parsley, chopped

Dissolve bouillon cubes in boiling water. Add the juice drained from the red beets. Add 1 cup chopped beets; save the rest of the beets for another day.

Rinse the beans in a colander. Add to the soup, bring to boil. Add lemon juice and seasonings. Remove from the heat; top with sour cream and parsley.

SERVES 6.

French Provincial Onion Soup

Paris may favor fancy foods, but the core of the French nation, all those who live in small provincial towns and villages, prefer simple, hearty eating. One of the best old-fashioned French dishes is the aromatic, rich brown onion soup. This is one case when you will get best results using a home-made, strong beef stock, but canned beef bouillon can be substituted. When properly cooked and served in an original French way, onion soup is a real delicacy, perfect for the first course at even the most elegant dinners. It may be served with the onions or strained.

2 quarts beef bouillon
6 cups onions, thinly sliced
3 tablespoons shortening
½ teaspoon sugar
3 tablespoons instant flour
1 cup red table wine
Salt and pepper
12 slices French bread
1 cup Swiss cheese, coarsely grated

Heat the bouillon. Fry the onions very slowly in hot shortening until tender. Add sugar and flour; fry, stirring constantly until golden brown. Slowly stir in the hot bouillon and blend evenly. Add wine; salt and pepper to taste. Simmer covered for 20 minutes.

Sprinkle the bread with cheese, place on a cookie sheet in the middle of the oven. Broil until the cheese melts. Place each piece of bread in a soup bowl. Pour the soup over it and serve immediately.

SERVES 12.

Cabbage Soup from Finland

Everywhere in the world, there are plenty of free-loaders who love to accept invitations, but seldom want to reciprocate. They always find an excuse: "Oh, I'm really ashamed of myself! I should have so many people over, but somehow I'm always too busy." We never lack time for what we really want to do, however. It isn't necessary to entertain in an elaborate way. An occasional simple and tasty family dinner will make you more friends than a cocktail party for a hundred once a year. This is an excellent, simple first course for just such an informal dinner party.

½ medium head cabbage, shredded
2 cups boiling water
Salt
3 tablespoons instant flour
1 ½ quarts milk
1 tablespoon butter
1 teaspoon sugar

Cook cabbage in salted water for 15 minutes. Mix flour with a little cold milk; add to the remaining milk. Add milk to cabbage and simmer for 10 minutes. Season with butter, salt, and sugar.

SERVES 8.

Hungarian Sauerkraut Soup

Sauerkraut is used in Central Europe in many ways. It may be served raw with apples and sugar as a salad. It is often stewed with various meats, sausage, and mushrooms. Another good way of using it is to cook a tasty, nourishing sauerkraut soup.

2 pounds sauerkraut, drained
8 cups water
2 pounds pork ribs, separated
6 slices bacon, coarsely chopped
1 onion, chopped
½ pound sausage, sliced
1 teaspoon dill weed
1 teaspoon paprika
Dash of garlic
1 tablespoon flour
¼ cup sour cream

Cook sauerkraut with pork ribs until the meat is tender. Fry the bacon with the onions until golden. Add to the soup. Add sausage, dill weed, paprika, garlic, and flour mixed with sour cream. Bring to a boil.

SERVES 8.

Cream of Carrot Soup from Finland

The Finns are the heroic people of a small country that defended itself successfully during the winter of 1940 against the aggression of the Soviet Union. The Russians attacked with all their mighty hardware. But what can you do with huge, heavy tanks in dense Finnish forests covered with deep snow? When the darkness had covered battlefields, Finnish farmers treated their weary soldiers with nourishing soup.

5 carrots, scraped
1 cup boiling water
½ teaspoon salt
2 tablespoons butter
3 tablespoons instant flour
1½ quarts milk, boiling
3 tablespoons green parsley, chopped
Croutons

Cook carrots in salted water for 15 minutes. Whip in a blender. Heat the butter in a saucepan, add flour. Fry a few minutes, stirring. Add milk in a thin stream, stirring. Bring to a boil and simmer uncovered for 10 minutes. Add the carrot purée and parsley. Season. Serve with croutons.

SERVES 8.

Scandinavian Summer Soup

In Scandinavia there is no such thing as dinner without soup. This soup, a favorite in Sweden, is extremely easy to make, and refreshingly different.

1 carrot, sliced
½ small cauliflower, coarsely shredded
1 cup potatoes, thickly sliced
½ cup frozen peas
1 cup string beans, sliced
2 cups boiling water
Salt to taste
1½ tablespoons instant flour
1½ quarts milk
½ cup spinach, shredded
1½ tablespoons butter
1½ tablespoons green parsley, chopped

Cook carrots, cauliflower, potatoes, peas, and beans in salted, boiling water for 15 minutes. Combine the flour with a little cold milk. Stir in the remaining milk, add to the soup. Add spinach; cook for 10 minutes. Add butter, season with salt, add parsley.

SERVES 8.

Hungarian Peasants' Bread Soup

Country cookery is often quite imaginative. This is soup to cook when you learn at the last moment that unexpected company is coming to dinner, and there just is not enough time to run to a supermarket. Bread soup is done in a jiffy, and it is certainly tasty enough to bring you compliments.

6 slices bacon, diced
6 slices bread (the older, the better), diced
2 onions, chopped
1 tablespoon green parsley, finely chopped
6 cups water
Salt to taste
1 tablespoon paprika
3 eggs, beaten

Fry the bacon slowly until transparent. Add the bread, onions, and parsley. Fry until golden. Put everything into a pot of boiling water. Season with salt and paprika. Add the eggs in a thin stream, when water is boiling vigorously. Serve immediately.

SERVES 6.

Spinach Soup from Oslo

The Norwegians are most practical people. They believe that a good soup should be filling and also a healthful, fundamental part of every dinner.

Spinach soup effectively takes the place of an appetizer and requires less effort. This is another recipe that can be cooked in a jiffy. Prepare it a day ahead of your party and reheat, or if it happens to be a hot day, serve the soup chilled because it tastes even better that way. Fill the soup bowls in the kitchen and place them on the table before you call your guests to dinner.

6 hard-boiled eggs, quartered
1½ quarts canned chicken bouillon
4 tablespoons instant flour
⅔ cup milk
2 jars strained spinach (baby food)
2 tablespoons lemon juice
1 egg yolk
1½ tablespoons soft butter
Salt to taste

Prepare the eggs. Heat the bouillon to boiling. Mix the flour with milk, adding it by spoonfuls and mixing. Add to the bouillon; bring to a boil. Remove from heat. Add spinach and lemon juice, then the egg yolk mixed with butter. Season with salt. Place 1 egg in each soup bowl.

SERVES 6.

Lemon Soup from Warsaw

I suspect that the principal cause of the comparative un-popularity of soups in America is that starting one from scratch seems so time consuming and the mediocrity of canned ones is not too tempting. In most European countries soup is a must. One can easily skip appetizers, but in Warsaw, Berlin, or Budapest, a family dinner cannot be considered complete without a tasty soup.

It takes 10 minutes to prepare this gourmet specialty. Serve it hot in the winter time or well-chilled on hot summer days.

3 cups cooked minute rice
7 cups canned chicken broth
3 tablespoons instant flour
2 tablespoons soft butter
1 lemon, peeled and sliced very thin
2 tablespoons lemon juice
½ cup sour cream

Prepare the rice. Bring the broth to boil.

Mix the flour with butter, add a few spoons of hot broth, mixing. Add mixture to the broth, bring to boil, add the rice, and remove from heat. Add the lemon, the juice and sour cream.

SERVES 8.

Goulash Soup from the Hungarian Countryside

In Hungarian peasants' cottages one can always see bunches of paprika hung up to dry. Paprika is the national vegetable and spice of Hungarians. In its fresh, green state, sweet and juicy, or ripe red, dried out and powdered, it can be found in almost every dish. The main midday meal in the countryside often consists only of rich goulash soup followed by pancakes.

2 medium onions, finely chopped
3 tablespoons lard or shortening
1 tablespoon paprika
2 pounds stewing beef chunks, trimmed and cut into
*   1-inch cubes*
8 cups water
Salt to taste
1 green pepper, sliced
1 tomato
2 carrots, sliced
2 parsley roots or parsnips, sliced
2 potatoes, cut into cubes
2 cups cooked noodles

Fry the onions in hot fat until golden; add paprika, meat, 1 cup water and salt to taste. Simmer for 1½ hours.

Add the vegetables and 7 cups of water. Salt to taste. Cover and simmer ½ hour. Add noodles and serve.

SERVES 10.

Polish Blueberry Soup

Among the endless variety of dishes in Polish cuisine—influenced by French and Italian queens, by wives of Polish counts and gentry, by invading armies of greedy neighbors, as much as by folk traditions—the most original are Polish soups. And the best are the fruit soups. Served cold on hot summer days, they are deliciously refreshing and just marvelous.

Fruit soups are not overly sweet and people in Poland eat them as a first dinner course with croutons or noodles for a more hearty meal. Or sometimes, they are served after the entrée and that way the dessert may be skipped.

1 quart blueberries
1 slice white bread
4 cups water
½ teaspoon cinnamon
¼ teaspoon cloves
½ cup sugar
⅔ cup sour cream

Add the blueberries and bread to 1 cup boiling water. Bring to boil, reduce the heat, cover, and simmer for 5 minutes.

Put into a blender and mix for a few seconds. Pour into a serving dish, add 3 cups of boiling water, spices, and sugar; mix. Refrigerate for several hours.

Add sour cream to the soup tureen and mix in before serving. Serve croutons on a separate plate.

SERVES 6.

. . . .

Polish Plum Soup

Other fruit soups are prepared the same way as blueberry soup. One of the best is plum soup. In the recipe above you may substitute 1½ pounds pitted prune plums, or 3 pints pitted cherries for the blueberries. With plums and cherries many people like a little more sour cream, ¾ cup instead of the ⅔ cup used with the blueberry soup. Plum or cherry soups are so colorful that some people feel that it is a shame to eat them.

. . . .

Polish Strawberry Soup

This is another summer delight and a real gourmet treat that can be prepared with no more effort than the opening of a can. Thanks to modern technology, we can forget the tedious job of rubbing the berries through a sieve. This push-button recipe gives us neither a saucepan to wash nor the worry of killing precious vitamins.

1 quart strawberries, washed and hulled
1 quart buttermilk
½ cup sugar
⅔ cup sour cream

Put the strawberries into a blender with 1 cup buttermilk. Mix for a few seconds. Pour into a soup tureen. Add sugar, mix well. Add the rest of the buttermilk mixed with sour cream. Chill thoroughly. Serve with home-made croutons.

SERVES 6.

. . . .

Scandinavian Dried Fruit Soup

With the good supply of fresh, canned, and frozen fruit available in every supermarket, we seldom used dried ones. They can, however, provide a welcome diversity to our menus. Use this old Scandinavian recipe for a gourmet dinner and you may be sure of a pleasantly unusual effect.

> 2 cups mixed dried fruit, sliced
> ⅓ cup golden raisins
> 1 ½ quarts water
> ¾cup cranberry juice
> Dash salt
> ½ teaspoon cinnamon
> ¼ cup minute tapioca
> 1 cup sugar

TOPPING

> 1 cup whipping cream

Soak dried fruit and raisins in water overnight. In the morning add remaining ingredients and simmer for 1 hour. Chill.

Pour into individual soup bowls. Place a heaping spoon of whipped cream into each bowl.

SERVES 8.

. . . .

Jewish Matzo Dumplings

This is a traditional Passover dish, cooked by Jewish mothers and served in soups at Passover feasts. Passover, the eight days of the Jewish spring festival, is primarily a family holiday. Symbolic foods relevant to Jewish history are served, one of the principal being the matzo—bread made without any leaven. Matzo, matzo meal, or matzo flour are the ingredients found in many Passover dishes. It can be purchased in Jewish gourmet shops and in most larger supermarkets in this country.

4 eggs
½ cup water
⅓ cup cooking fat
¾ teaspoon salt
⅛ teaspoon pepper
1 cup matzo meal

Beat the eggs with water and melted fat with a fork for 1 minute. Add salt, pepper, and matzo meal. Mix well, and let stand for 25 minutes.

Form small balls and drop them carefully into a large kettle with boiling, salted water. Cook for 20 minutes. Transfer to a hot soup with a slotted spoon.

SERVES 5.

. . . .

French Croutons

Croutons originated in refined French restaurants and were quickly adopted by most European countries. They are tiny bits of toast that go well with many creamy or clear soups.

Croutons are best when fresh, but they stay crisp when kept uncovered overnight. Serve them in a separate dish and pass around the table. Each guest sprinkles croutons directly into his soup bowl.

Croutons are sold in many supermarkets, but they could not even compare with those you can easily make at home.

6 slices white bread
4 tablespoons soft butter

Spread bread slices with butter on both sides. Pile them one upon another. Cut with a thin, sharp knife into small even cubes. Sprinkle them on a cookie sheet.

Bake in a hot 450° oven for 10 minutes or until golden. Turn after the first 5 minutes.

SERVES 6.

Stuffed Pike from Finland

In Finland Santa Claus never sneaks in through a chimney. He comes right through the door early in the evening on Christmas Eve, to be touched and admired by everyone. Sometimes he even stays with the family for a whole hour, talking and singing Christmas carols. Finnish boy scouts often earn money by playing the role of Santa and visiting homes where there are children.

Here is one of Finland's typical Christmas Eve fish dishes.

STUFFING
½ cup rice
½ pound spinach
2 eggs, slightly beaten
Salt to taste

One 3-pound pike
2 tablespoons butter
Salt to taste
¼ cup bread crumbs
½ cup boiling water
Green parsley sprigs

TO MAKE STUFFING: Cook the rice and spinach the usual way. Combine rice and spinach with eggs, season with salt.

TO BAKE FISH: Stuff the fish cavity and sew it up. Brown the butter in the oven in a baking dish. Put the fish in and baste with butter. Sprinkle with the bread crumbs. Bake in 350° oven for 40 minutes. When half done, add water. Baste a few times.

Cut into 6 portions in the baking dish. Garnish with parsley and serve.

SERVES 6.

· ·

Swedish Fish in Horseradish

Gourmet cooks feel that filleted fish is never as tasty as it can be when prepared whole, bones and all. It requires skill, however, to separate the meat from the bones while eating. It is important always to move the fork in the same directions the bones go in order to avoid detaching them from the vertebrae. In Europe special knives are usually served with fish dishes. They are dull and are used to hold, not to cut, the fish.

One 2-pound fresh whitefish, pike, or bass
1 large onion, quartered
4 peppercorns
Salt to taste
3 tablespoons butter
3 tablespoons instantized flour
3 tablespoons prepared horseradish
2 hard-boiled eggs, coarsely chopped
2 tablespoons green parsley, chopped

Place the fish in a kettle. Cover with water and add the onion, peppercorns, and salt. Simmer for 20 minutes.

Heat the butter, add the flour, and fry at medium heat for a few minutes, stirring. Gradually add enough fish stock to make a thick sauce. Simmer for 5 minutes. Add horseradish.

Transfer the fish carefully onto a warmed serving platter. Pour the sauce over the fish. Sprinkle with the egg and parsley.

SERVES 6.

Hungarian Fish Dinner

The easiest thing for a young hostess to serve is a whole elegant and savory dinner in one dish. Hungarians have many ideas on the matter. Let's follow their example, cook in advance, and have fun with the guests. Serve with tomato salad, hot rolls, and rosé wine.

6 portions mashed potato
1 egg, beaten
¼ teaspoon nutmeg
2 pounds spinach
3 pounds fish fillets
3 tablespoons butter

SAUCE BECHAMEL
3 tablespoons butter, melted
4 ½ tablespoons instantized flour
2 ¼ cups milk
Salt to taste
¾ teaspoon lemon juice
3 egg yolks
¼ cup grated Parmesan cheese
¼ cup bread crumbs

IN THE AFTERNOON: Mix mashed potatoes with the egg and the nutmeg. Spoon the mixture around the edges of a large buttered baking dish. Boil spinach in a large kettle with salted water for 5 minutes; drain. Arrange spinach leaves in the middle of the potato ring.

Divide fish fillets into 6 portions. Fry in hot butter on both sides until golden. Place on spinach leaves.

TO PREPARE THE SAUCE: Mix the butter with the flour over low heat. Gradually stir in the milk. Bring to a boil, stirring constantly. Remove from the heat. Season with salt and lemon juice. Stir a few tablespoons of the sauce into the egg yolks, then combine.

Pour the sauce over the fish and mashed potatoes. Sprinkle with cheese and bread crumbs. Cover and refrigerate.

BEFORE SERVING: Take the dish from refrigerator 1 hour before baking. Bake uncovered in a 400° oven for 15 minutes or until golden.

SERVES 6.

Baked Fish from Lake Balaton

Hungary's Lake Balaton abounds in fish of good quality. The soft water and velvety sands make an easy life for the fish, and their flesh is white and tender. Some specimens weigh up to 20 pounds. Hungarians prepare their fish carefully. As in most of their dishes, green pepper, paprika, and sour cream help give it a specific national flavor.

This is a complete dinner in one dish. Serve with garlic bread, sweet butter, and white wine.

3 pounds fish fillets
Salt and paprika
6 medium potatoes, parboiled, sliced
3 strips bacon, cut in halves
3 onions, sliced
3 green peppers, sliced
3 tomatoes, sliced
2 tablespoons butter, melted
½ cup sour cream
1 teaspoon flour

Cut the fish fillets into 6 portions, sprinkle with salt and paprika.

Arrange potatoes in the bottom of a large baking dish which has been buttered. Arrange fish portions over potatoes. Place a piece of bacon on each piece of fish. Cover with onions, green peppers, and tomatoes. Sprinkle with butter. Bake in 350° oven for 20 minutes. Mix sour cream with flour; pour over the dish. Bake another 20 minutes.

SERVES 6.

Carp Jewish Style

Carp is the favorite fish of East European Jews. In that part of the world it is customary to breed carp in small artificial ponds. When the fish reach an average size of approximately two pounds, the water is drained and all the carp are sent to the market.

The meat of European carp is very delicate and tasty. It is considered one of the most popular fish available. Our American carp caught in bays are not as good, but can make a nice dish if prepared carefully.

2 pounds carp
Salt and pepper
½ teaspoon cloves
2 tablespoons vinegar
3 tablespoons butter
1 large onion, sliced
½ cup beer
2 ounces raisins
½ teaspoon grated lemon rind

Cut the carp into 1-inch-wide portions. Sprinkle with salt, pepper, cloves, and vinegar.

Heat the butter and fry the onions till golden. Add the beer; heat. Place the fish in the beer sauce, cover, and simmer for 30 minutes.

Remove the carp to a warmed platter. Add the raisins and lemon rind to the sauce. Cook uncovered on a high heat for a few minutes. Pour over the carp. Serve with noodles.

SERVES 5.

Braised Fish Fillets from Norway

Women from Norway, the country with thousands of fiords, know how to serve fish. My Norwegian cousin has sent me this recipe for a small intimate dinner—simple, inexpensive, and low in calories.

Serve with boiled potatoes.

6 tablespoons butter
1 ½ pounds fish fillets, barely defrosted
Salt and pepper
1 cucumber, peeled, sliced
4 tomatoes, sliced
1 onion, sliced
1 green pepper, sliced
¼ cup strong bouillon from cube
3 drops mint flavoring

Heat (but don't brown) 4 tablespoons butter. Place fish fillets in melted butter. Sprinkle with salt and pepper. Cover with vegetables. Pour bouillon mixed with mint over the fish. Dot with remaining butter. Cover and simmer for 20 minutes.

SERVES 4.

· ·

Jewish-Style Fish in Cream Sauce

Jews developed special skills in the preparation of fish dishes because of their many religious prohibitions relating to cooking and eating meats. Fish dishes also suited the practical Jewish approach to life. Fish is less costly than most meats, low in calories, and a real time-saver. It is also a perfect source of protein, essential vitamins, and minerals.

This savory dish is easy to prepare and deliciously different.

1 stalk celery, diced
1 tablespoon green parsley, chopped
1 large onion, sliced
3 pounds fish fillets
Water to cover
1 teaspoon salt
1 tablespoon lemon juice
1 cup coffee cream
1 tablespoon butter
1 tablespoon instantized flour
1 tablespoon milk
¼ teaspoon salt
Dash of pepper

Line the bottom of a large pot with vegetables. Arrange fish fillets over them, cover with water, and add salt and lemon juice. Cover, bring to boil, and simmer for 20 minutes.

Heat the cream with butter. Blend the flour with milk; add a few spoonsful of hot cream; mix. Add to the rest of the cream and bring to boil. Cook at low heat for a few minutes, stirring constantly until smooth and thick. Season. Remove the fish carefully to a warmed platter, pour the sauce over it and decorate with green parsley. Serve with boiled potatoes, carrots, and peas.

SERVES 8.

Danish Fish with Cheese

When a Norwegian or a Swede becomes bored with things at home or perhaps a little depressed, he takes a few days vacation in Copenhagen. Denmark is the swingingest of Scandinavian countries. It is a land of happy, smiling people, beautiful architecture, and an abundance of excellent restaurants.

One Danish favorite is this broiled fish, which can be prepared in 15 minutes and may even impress your mother-in-law. Broil it in an attractive dish, serve with red cabbage, and boiled potatoes sprinkled with butter.

3 pounds flounder fillets
1 teaspoon salt
½ teaspoon pepper
¼ pound butter
1 cup crumbled blue cheese
2 tablespoons lemon juice
1 tablespoon chopped green parsley

Wash flounder fillets; dry with paper towels. Place in a buttered baking dish. Sprinkle with salt and pepper.

Melt the butter in a small skillet. Stir in the cheese and lemon juice. Spoon half of the sauce over fish. Place under preheated broiler. Broil for 3 minutes. Spoon the rest of the sauce over fish. Broil it for 3 more minutes or until golden. Sprinkle with parsley.

SERVES 8.

Roast Chicken Polish Style

Poultry has always been considered a special treat in Europe. A typical Sunday dinner at a Polish summer resort consists of a cold fruit soup with croutons or noodles, roast spring chicken with a special Polish stuffing, cucumbers in sour cream, young carrots, tiny young potatoes (the size of walnuts) boiled and sprinkled generously with melted butter and dill weed. Vanilla ice-cream is the traditional dessert for this meal.

Two 2–3-pound chickens
Salt
¼ pound butter, melted

STUFFING
2 chicken livers, finely chopped
1 ½ cups toasted bread crumbs
2 eggs, slightly beaten
¼ pound butter, melted
Dash pepper
Salt to taste
1 tablespoon dill weed
½ cup milk

TO MAKE STUFFING: Combine liver with bread crumbs, butter, eggs, and spices. Add as much milk as needed for a sour-cream consistency.

TO ROAST CHICKENS: Sprinkle chickens with salt. Stuff. Roast in 350° oven for 2 hours, basting often with butter and pan juices. Cut into quarters.

SERVES 8.

. . . .

Chicken in Lemon Sauce from Warsaw

This is my grandfather's favorite dish; for years it could easily put him in a good mood. I am afraid, however, that grandmother used it for that purpose too often because in time he became suspicious. Whenever he saw chicken in lemon sauce on the table, he expected bad news to follow.

Serve this delectable lemony chicken with rice, carrots, and a lettuce salad.

One 3-pound chicken, boiled, skinned, boned
1 tablespoon green parsley chopped

SAUCE
2 tablespoons butter, soft
3 tablespoons instant flour
1 ¾ cups chicken broth, boiling
2 teaspoons lemon juice

TO MAKE THE SAUCE: Mix the butter with the flour well. Stir in a few tablespoons of boiling broth. Add to the remaining broth, bring to a boil. Add lemon juice.

Add the chicken to the sauce, heat. Transfer to a warmed serving dish. Sprinkle with parsley.

SERVES 6.

* * *

Turkish Chicken with Rice

Turkish wives like to cook the entire dinner in one dish. Some people claim that it is because they are a little lazy and don't want to scrub too many pots and pans. But I think that they feel dinner tastes better that way. Let's find out who is right.

This delicious chicken can be prepared and cooked just in 1 hour.

One 2½-pound chicken cut up into small portions
3 tablespoons oil
3 medium onions, chopped
1 bay leaf
10 ounces long-grain rice
3 tomatoes, sliced
1 green pepper, sliced
10 peppercorns
½ cup golden raisins
2 cups chicken bouillon from cubes, boiling
Pinch saffron
Salt to taste
3 ounces black olives
1 cup canned tomato sauce

Sauté chicken in 2 tablespoons oil in a deep frying pan on medium high heat until pale golden. Add the remaining oil, the onions and bay leaf. Fry, stirring until the onions are transparent. Add rice; fry for 5 minutes, stirring. Add tomatoes, green pepper, peppercorns, raisins, and bouillon mixed with saffron. Season with salt. Cover and place in 450° oven for 25 minutes. Add olives, stir. Cover with tomato sauce and let stand uncovered in the oven for another 5 minutes.

SERVES 6.

Jewish Chicken with Grapes

The most unifying and universally observed of all Jewish ceremonies is Passover. When Jews sit down to the Passover meal, they recount the 3,000-year-old story of the Angel of Death "passing over" the children of the Israelites when God smote the first-born of the Egyptians in order to induce the Pharaoh to let the Jews go.

Chicken, especially with fruit dressings, is a favorite dish of Passover dinners.

3 tablespoons matzo meal
1 teaspoon salt
¼ teaspoon pepper
4 pounds chicken breast and legs
¼ cup oil
⅓ cup orange juice
½ cup dry white wine
2 tablespoons honey
1 ½ tablespoons green parsley, chopped
Rind of one orange, coarsely grated
1 cup seedless grapes, cut into halves
1 orange, peeled, thinly sliced

Combine matzo meal, salt, and pepper. Sprinkle chicken parts. Heat the oil and fry chicken until golden. Add orange juice, wine, honey, and parsley. Simmer covered for 45 minutes. Add orange rind and simmer for 10 minutes more.

Remove chicken into a warmed serving platter. Add grapes and orange slices to the sauce and heat. Pour over the chicken and serve.

SERVES 6.

Spanish Chicken with Vegetables

Are you tired of fried chicken? If so, it is time to try something new. This Spanish dish makes a whole meal in one deep frying pan. It is especially suitable for a Sunday dinner, for entertaining in-laws or other relatives—when you want to serve something tasty, but not too expensive.

Serve with tomato salad and hot garlic bread.

6 portions chicken, breasts and legs
¼ teaspoon garlic
1 tablespoon paprika
¼ teaspoon pepper
⅓ teaspoon salt
5 tablespoons oil
2 large onions, cut into thick slices
1 bay leaf
½ cup white table wine
1 cup boiling water
6 medium potatoes, diced
2 green peppers, sliced
1 cup canned peas, drained

Combine garlic with paprika, pepper and salt. Sprinkle chicken breasts and legs. Refrigerate overnight.

Sauté onions in hot oil until golden. Remove the onions. Fry chicken until golden. Add to the frying pan: onions, bay leaf, wine, and water. Cover and simmer for 1 hour. Add potatoes and green peppers. Simmer ½ hour more. Add peas. Heat and serve.

SERVES 6.

Hungarian Chicken with Paprika

Chicken prepared this way makes an inexpensive Sunday dinner, easy to prepare. It can be cooked in the morning or even the previous day and reheated. Add sour cream at the last moment when your guests are already seated. Transfer into a warmed deep serving dish and pass around.

Chicken with paprika calls for real Hungarian butter dumplings. But if you are pressed for time, rice will do. For an attractive colorful table a green vegetable—perhaps peas or Brussels sprouts—is a must. A bowl of salad will supplement this meal nicely.

One 2-pound chicken, cut up
1 medium onion, chopped
4 tablespoons lard
1 tablespoon paprika
2 tablespoons tomato purée
¾ teaspoon salt
¼ cup water
¾ cup sour cream

Wash the chicken and dry with a paper towel.

Fry the onions in a large skillet in hot lard till pale golden, then add paprika. Add chicken pieces, fry on both sides till golden.

Add tomato purée, sprinkle with salt. Add water, cover and simmer for 1 hour.

Then add sour cream, heat but not to the point of boiling, otherwise the cream will curdle.

SERVES 4.

Roast Pheasant or Grouse the Basque Way

To many, hunting is an exciting sport. For us, in the United States, it means spending some time close to nature and proving our skills. But in the country of the Basques, it was often a matter of life or death. For centuries, the Pyrenees Mountains provided excellent hiding places for all kinds of outlaws and revolutionaries who depended on game birds for survival. And Basque shepherds spending month after month on distant, lonely pastures have found pheasant or grouse a happy diversion from their monotonous diet. They have invented an original way of roasting their catch.

1 pheasant or grouse
½ cup brandy
½ teaspoon salt
5 parsley sprigs
¼ pound butter, melted
8 small potatoes, peeled

Wash the bird thoroughly and dry. Rinse inside and out with brandy, reserving 4 tablespoons. Sprinkle the body cavity with salt and parsley. Truss. Place your bird in a baking pan, arrange potatoes around the bird. Pour over the butter. Roast at 425° for 50 minutes, basting frequently.

Transfer the bird and potatoes to a warmed serving platter. Remove the trussing. Add the saved brandy to the drippings in the pan and flame. Pour over the bird. Serve with cranberries or currant jelly.

SERVES 4.

Jewish Liver with Oranges

My mother loved to buy fabrics for her dresses in a little Jewish yard goods store not too far from our home in Warsaw. It always took her a long time to make her selection, and in addition she usually stopped at the apartment in back of the store for a lengthy chat with the owner's wife. Knowing that such afternoons tend to be boring to little girls, the good woman offered me all kind of goodies. Once we stayed for dinner and had liver prepared in this original way. It takes 15 minutes to prepare and only 10 minutes to cook.

6 generous slices beef liver, well trimmed
¼ pound shortening
Salt and pepper to taste
Juice of one orange
2 oranges, peeled, sliced

Sauté liver in ¾ of the shortening over medium heat until it is still pink inside and golden but not dark or hard on the outside. Transfer liver onto a warmed serving platter, sprinkle with salt and pepper. Place the platter on a warmer or on top of a saucepan with boiling water.

Pour the orange juice into the frying pan, stir, and scratch off all the pan juices to mix them well with the orange juice. Season with salt and pepper. Bring to a boil and add orange slices. Heat while shaking the pan to prevent sticking. Garnish the liver with hot orange slices. Quickly add the remaining shortening to the sauce, beat with the fork to speed up melting, and pour over the liver.

SERVES 6.

* *

Roast Veal the Russian Way

European veal is younger and whiter than American, but it is possible to buy a good roast of veal in this country. Careful preparation is needed for it to be tender and juicy. Aromatic seasoning and a good sauce greatly enhance the flavor of veal.

It can make a savory party dinner and a nice variation on our more common roast beef. It is a convenient recipe, with most of the cooking done well ahead of the party.

Serve with rice or noodles, spinach or asparagus, and tomato or red cabbage salad.

4 pounds boneless roast of veal
1 cup oil
2 carrots, sliced
1 large onion, sliced
1 bay leaf
3 tablespoons butter
Salt
1 teaspoon parsley, chopped
One strip fresh pork fat
*Mushroom Sauce**

IN THE AFTERNOON: Wash the roast and thoroughly dry. Fry in hot oil over medium heat until golden brown on all sides. Simmer the carrots, onions, and bay leaf with butter in a large casserole for 10 minutes. Sprinkle the roast with salt and parsley, and place it on the vegetables. Cover with pork fat. Place a sheet of aluminum foil over the roast. Cover the casserole and roast in 325° oven for 1½ hours.

Remove from oven and let stand for ½ hour, then cut into ¼-inch thick slices. Strain and degrease the pan juices. Place the meat in a buttered baking dish. Sprinkle with the juices, cover with mushroom sauce. Refrigerate covered.

½ HOUR BEFORE SERVING: Place uncovered in oven preheated to 375° until golden brown.

SERVES 8.

. . . .

Veal Dinner from Zürich

Zürich is a very old city which was ruled by the Romans more than 2,000 years ago. In the fourteenth century, it became a part of Switzerland. There are many charming old buildings in Zürich, including museums and churches that have been there for hundreds of years. There is also a modern section with comfortable contemporary homes.

The German part of Switzerland is famous for its cuisine. Veal is one of its specialties. The following recipe shows the original way of preparing and serving it in Swiss homes. One of my friends from Zürich was kind enough to share her mother's notes with me. This is not an inexpensive dish, but when you want to entertain special guests in style without too much trouble, it would be the perfect choice.

Serve with sautéed mushrooms, grilled tomato halves, green beans, and rice or noodles.

1 medium onion, thinly sliced
4 tablespoons butter
2 pounds veal cut for scallopini (thin strips)
4 teaspoons flour, instantized
1 cup dry white wine
1 cup chicken consommé from a cube
½ cup whipping cream
Salt and pepper
1 tablespoon green parsley, finely chopped

Sauté onions in hot butter until lightly yellow. Add the meat; fry for a few minutes until it gets white. Sprinkle with flour; stir. Immediately stir in wine. Bring to a boil. Add consommé and simmer for 3 minutes. Add cream, warm up, but do not boil. Season with salt and pepper. Transfer to a warmed serving dish, sprinkle with parsley.

SERVES 6.

. . . .

Veal Schnitzel from Vienna

Wiener Schnitzel is simple to prepare and a nice choice for a small intimate dinner for two or four. Use a heavy 9-inch skillet for two veal steaks. If you need more, use two skillets. Don't crowd your meat.
Serve with spinach and mashed potatoes sprinkled with parsley.
In Central Europe people drink beer with their meals rather than before or after. Serve beer with your schnitzel. Rosé wine could also complement your dinner nicely.

10 to 12 ounces veal cutlets
3 small eggs
⅓ cup bread crumbs
Salt and pepper
3 tablespoons bacon drippings or shortening
1 tablespoon butter
2 slices lemon

Trim your meat thoroughly, cut into 2 serving pieces. Pound well. Form rounded steaks. Dip in 1 beaten egg, then in bread crumbs. Sprinkle with salt and pepper.
Heat bacon drippings in a skillet. Fry veal steaks in medium high heat until golden brown on one side. Turn, fry till golden. Add butter, turn the heat to low, and fry for 5 minutes more.
Fry 2 eggs sunny side up.

Place one egg on each schnitzel, garnish with lemon slices.

SERVES 2.

. . . .

Hungarian Veal and Paprika

In Hungary good food combines naturally with music. And in most cases the music is that of a gypsy violin alternately crying or laughing, making people happy and sentimental at the same time. Good background music helps every dinner tremendously. Somehow food always tastes better when music puts you in the right mood.

Serve Hungarian veal with egg barley or egg noodles, and . . . gypsy melodies.

2 large onions, chopped
4 tablespoons bacon drippings or butter
4 pounds shoulder veal, cubed
1 ½ tablespoons paprika
¼ cup water
1 ½ teaspoons salt
1 cup sour cream

Fry the onions in hot fat until golden. Add veal and paprika. Brown the meat; add water and salt. Simmer covered at low heat for 1 hour. Add sour cream, heat, but do not boil.

SERVES 8.

. . . .

Pork Chops from Poland

Pork is the favorite meat in Poland, and the most common way to serve it is in the form of pork chops. They must be very well trimmed, tender, and fried to a glorious golden-brown crust.
Serve with cabbage or stewed sauerkraut and potatoes, boiled or mashed.

6 medium, lean pork chops from center-cut pork loin
6 tablespoons flour
2 large eggs, lightly beaten
Salt and pepper
1 cup bread crumbs
½ cup bacon drippings or shortening

Trim all the fat off. Pound the meat well on both sides. Roll in the flour. Dip in the egg mixed with salt and pepper. Roll in bread crumbs, and press them to the surface.
Fry in hot drippings, 2 or 3 at a time on both sides on medium heat. Place in a large baking dish. Place in a 325° oven for 15 minutes.

SERVES 6.

Roast Pork Loin from Moravia

There isn't much to do on Sundays in small Moravian towns. Dinner is the most important event of the day, and the favorite Sunday dinner in Moravia is roast pork loin with cabbage and potatoes. Whole families, grandparents, aunts, and cousins, gather around the table for a long, relaxed meal and the latest gossip. The roast must be juicy and tender. Moravian cooks keep it for several hours in a slow oven. Ask the butcher to cut the bone lengthwise for easier slicing.

4 to 6 pounds pork loin
Salt
1 tablespoon caraway seeds
1 large onion, sliced

Score the fat of the loin in small squares. Sprinkle the meat with salt and caraway seeds. Place in a roasting pan, fat-side-up. Roast in 325° oven counting 40 minutes per pound. Add the onion when the meat is half done. When ready, turn off the heat and leave the roast in the oven for 15 more minutes.

SERVES 2 PER POUND OF PORK.

. . . .

Lamb in Cabbage from Norway

This is *Fårikål,* a Norwegian national dish, and the best contribution of the little northern country to Continental culinary traditions. Scandinavians usually serve it during the winter.

This stew has to simmer for 2 hours, but the preparation is simple and takes only 15 minutes. Serve in a large round bowl with baked potatoes and rye bread.

2 tablespoons butter, soft
1 large head (5 pounds) cabbage, coarsely sliced
4 pounds lamb (ribs or shoulder), cut into 3-inch pieces
5 large carrots, peeled, sliced lengthwise
1 ½ tablespoons salt
2 tablespoons peppercorns
½ cup instant flour

Spread the bottom of a kettle with butter. Arrange cabbage, meat, and carrots in several layers, starting and ending with cabbage. Sprinkle each layer with salt, peppercorns and flour. Add enough boiling water to cover ¾ of the stew. Bring to a boil covered; simmer for 2 hours. If the sauce is too liquid, add some more flour.

SERVES 8.

· ·

Leg of Lamb from Spain

I met a delightful Spanish lady who told me that in her native Madrid when people wish to entertain their guests nicely, they most often serve a good mutton or lamb roast. The roast is placed on lettuce leaves, on a warmed, large platter and garnished with olives. It is always carved at the table. No special gravy is prepared because with proper cooking there should be enough natural juices for sprinkling each slice.

Lamb roast is usually served in Spain with parsleyed potatoes and a big bowl of tossed salad.

ROAST
4- to 5-pound leg of lamb

MARINADE
2 cloves garlic
1 teaspoon salt
¼ teaspoon oregano
½ teaspoon paprika
Pinch of tarragon
4 tablespoons oil

Crush the garlic, mix with all the ingredients of the marinade. Spread the marinade over the lamb. Let stand at room temperature for 4 hours.

Preheat the oven to 500°. Place the roast in the oven; set the thermostat to 325°. Roast for 22 minutes per pound for a pink meat, 25 minutes per pound for well done.

SERVES 8.

· ·

Hungarian Wooden Plate Dinner

In the olden days, after the nomadic Magyar tribes settled in the Great Plain of today's Hungary, a special treat served to honored guests was a variety of meats on a wooden plate. It makes an interesting dinner for a small group of 4 or 6. Serve with cooked, sliced, and browned potatoes, and Hungarian Salad in separate bowls.

1 pound sirloin
1 pound veal cutlets
1 pound veal livers
6 strips bacon

Cut each kind of meat into 6 portions. Broil everything. Arrange fried potatoes on each plate; pile the meat over it.

SERVES 6.

Hamburgers from Prague

Each country has its own version of the hamburger. In Prague hamburgers are very different from ours, and they are never considered an inferior dish. This recipe is a tasty idea, worth trying, particularly in these days of high meat prices.

Serve with carrots, mashed potatoes, and lettuce in sour cream dressing.

1 medium onion, quartered
3 slices white bread
1 egg
½ cup milk
1½ pounds ground beef (chuck)
½ teaspoon salt
⅛ teaspoon pepper
1 tablespoon green parsley, chopped
½ cup bread crumbs
3 tablespoons bacon drippings

Combine the onions, bread, egg, and milk in a blender. Add to the meat, mix well. Season with salt, pepper, and parsley.

Form 8 flat cutlets. Roll in bread crumbs. Fry in hot drippings on medium high heat on both sides until golden brown. Place in 325° oven for 15 minutes.

SERVES 8.

Hungarian Beef Dinner

This recipe makes a lovely, casual wintertime dinner for young informal company. It can be prepared in advance and reheated, or put on the stove just before your guests arrive. It certainly does not keep the hostess in the kitchen.

All you need to add is a large bowl of salad accompanied by some rye bread and butter. For a special touch serve red table wine at room temperature, and put candles on the table.

For cooking Hungarian beef dinner use the nicest heat-resistant casserole dish you have, and then serve in it.

2 pounds round steak, well trimmed
4 tablespoons lard or bacon drippings
Salt
2 onions, sliced
¼ teaspoon garlic powder
1 teaspoon caraway seeds
1 tablespoon paprika
2 red or green peppers, sliced
1 cup canned tomato purée
½ cup water
8 medium potatoes, sliced

Cut the meat into 8 portions, pound them till very thin. Fry in hot lard on both sides; sprinkle with salt. Transfer the meat into a casserole dish.

Fry the onions in the fat left in the skillet, transfer to the casserole. Add the seasoning, peppers, tomato purée, and water. Cover, bring to boil, reduce the heat and simmer for 1 hour. Arrange potato slices on the meat, cover, and simmer another 20 minutes.

SERVES 8.

. . . .

Norwegian Beef Stew

This is a thrifty dish that requires long cooking but not too much labor. It is a good choice for a family dinner after Christmas when you must be economical.

Serve with rice.

1 ½ pounds stewing beef, cut into 1-inch cubes
1 large onion, sliced
2 tablespoons margarine
2 cups strong beef bouillon from cubes
One 1-pound can chopped tomatoes
2 green peppers, sliced
1 cucumber, peeled, diced
Salt and pepper
1 tablespoon instant flour
½ cup sour cream

Fry beef with onions in hot fat until golden. Transfer to saucepan. Add hot bouillon, tomatoes, peppers, cucumbers. Season generously with salt and pepper. Simmer for 2 hours. Add sour cream combined with flour, bring to boil.

SERVES 5.

. . . .

Roast Beef from Prague

For a girl in Prague, cooking beef is a special test. She is expected to present her fiancé with a perfect roast before the wedding day. With this recipe she is certain to win his approval.

4 pounds eye of the round
½ pound butter, melted
1 onion, sliced
1 heart of celery, sliced
10 peppercorns
1 bay leaf
¼ teaspoon thyme
¼ teaspoon allspice
Salt
4 ounces bacon
2 tablespoons red table wine
½ cup boiling water
1 slice lemon rind

GRAVY
1 tablespoon instant flour
1 tablespoon cold water
¼ teaspoon sugar
1 teaspoon lemon juice
½ cup sour cream

Pour the butter over the meat. Cover and refrigerate for 2 days.

Spread a baking pan with butter from the meat. Arrange onions, celery, peppercorns, and bay leaf in the pan. Place the meat on it. Sprinkle with thyme, allspice, and salt. Cover with bacon. Add wine, water, and lemon rind. Roast at 325° for 2 hours and 40 minutes. Remove the meat and cut into thin slices.

TO MAKE GRAVY: Whip the sauce from the pan in a blender. Stir in flour mixed with water. Bring to a boil. Season with salt, sugar, and lemon juice. Remove from the heat. Add sour cream. Pour over the meat.

SERVES 12.

German Thrifty Beef Roast

Besides the famous "sauerbraten" that requires marinating, thrifty Germans have found a labor- and money-saving way of roasting less-expensive cuts of beef. These roasts must be well done, but they still melt in the mouth.

3 to 4 pounds lean chuck or rump roast
Salt and pepper
2 large onions, sliced

Pound the roast well on all sides. Sprinkle with salt and pepper. Place in the roasting pan in a slow 300° oven for 1½ hours. Baste and cover the roast with onions. Set the oven at 350°. Roast for another hour and 15 minutes, basting three times. Remove from the oven and wait 10 minutes before slicing to prevent the loss of juices.

SERVES 8.

Spaghetti from Bologna

This dish is Italian, but popular and loved as much in Sweden and Germany as in Bologna. It is a meal for young active people with hearty appetites who don't need to spend a fortune to have fun and good food.

Prepare the sauce in the morning and reheat before serving. Serve spaghetti with a big bowl of tossed salad in Sauce Vinaigrette* and of course—Chianti.

SAUCE
4 tablespoons oil
4 large onions, finely chopped
6 ounces bacon, chopped
2 pounds ground chuck
4 tablespoons butter
1 pound canned tomato purée
Salt and pepper
2 cups strong beef bouillon from cubes

2 pounds spaghetti
4 ounces Parmesan cheese

TO MAKE SAUCE: Heat the oil, fry the onions until transparent, add the bacon and fry for a few minutes. Add the meat, fry, stirring for 5 minutes. Add butter and tomato purée; season with salt and pepper. Add bouillon and bring to a boil. Then simmer for 10 minutes. The sauce should be thick.

COOK SPAGHETTI: drain. Arrange in a serving bowl in a heap with a big hollow. Pour the sauce into this hollow. Serve grated cheese in a separate dish.

SERVES 8.

. . . .

Hungarian Butter Dumplings

Preparing butter dumplings does not take much time and the result is certainly worth the effort. They are so much better than any ready-made noodles you can ever buy.

As dumplings have to cook fast it is important to use a large kettle with plenty of water and never to cook at one time more than one portion. These dumplings go well with all meats cooked in a sauce and make a perfect dish for a small dinner party.

3 ¼ cups flour
1 egg
¾ teaspoon salt
3 tablespoons butter, melted
1 ½ cups milk

Measure the flour into a bowl. Add the egg mixed with salt, butter, and milk. Mix with a tablespoon for a few minutes.

Drop small portions of the dough from a teaspoon into boiling water, dipping the spoon into the water each time. Cook for 5 minutes. Drain in a colander. Transfer into a warmed dish. Sprinkle with the sauce from the meat.

SERVES 6.

· ·

Hungarian Quick Dumplings

Through centuries of wars and hardships, Europeans learned never to waste food. Even in well-to-do families, throwing out bread always was considered a sin. There were too many hungry people and animals in the streets. Little children, dropping a roll on the floor in anger or carelessness, were ordered to pick it up immediately and kiss it, asking forgiveness. This symbol of respect was supposed to guard them from hunger. It was a superstition, but one with deep wisdom and a good lesson in thrift.

Stale rolls and white bread make very nice, refreshingly different dumplings. Serve them with pot roasts and stews.

4 slices white bread, cut into small cubes
4 strips bacon, diced
1 large onion, finely chopped
1 tablespoon green parsley, chopped
2 large eggs, beaten
⅓ cup milk
¾ cup flour
Salt to taste
Dash of pepper

Brown the cubes in a 400° oven for 10 minutes.

Fry the bacon with onions and parsley until golden. Add bread cubes, mix. Combine the eggs with milk and flour, season with salt and pepper, stir into the bread cubes. Mix well.

With wet hands, form small dumplings. Drop into boiling water; bring to boil. Cook until they float; drain in a colander.

SERVES 6.

- -

Swedish Mushroom Sauce

One of the favorite sauces in Scandinavia is this mushroom sauce served over meat loaves, pot roasts, poultry, fish, potatoes, cauliflower, noodles, in various casseroles, and many other dishes.

Serve with a colorful bowl of salad.

2 cups mushrooms, thinly sliced
1 large onion, chopped
3 tablespoons butter
2 tablespoons instant flour
1 ½ cups light cream (half'n half)
Salt and white pepper

Sauté mushrooms and onions in hot butter until lightly golden. Sprinkle with flour, stirring. Fry for a few minutes. Stir in the cream gradually. Cover and simmer for 5 minutes. Season.

YIELDS 2½ CUPS SAUCE.

- -

Sauce Hollandaise from Everywhere

Sauce Hollandaise, French in origin, has long been an international dish. It was traditionally served with fish and white meats or vegetables. Lately it has become fashionable to serve it with meat fondue.

1 cup beef bouillon from cubes
1 cup dry white wine
8 egg yolks
½ teaspoon salt

Heat the bouillon and wine to boiling in a saucepan. Beat the egg yolks for 5 minutes. Add the hot bouillon and wine mixture in a thin stream, still beating. Heat in a double boiler, stirring until thick. Do not boil. Season with salt and serve.

This sauce can be reheated carefully over a low flame.

YIELDS 2 CUPS.

- -

A French Masterpiece — Sauce Béarnaise

In great French restaurants, there is a separate chef for meats, a separate one for vegetables, another one for soups, another for desserts, and a different one for sauces. This last chef receives the highest pay of all because he is considered the most important. French gourmets believe that it is the sauce that makes the dish.

Outside of a simple gravy we are not accustomed to cooking special sauces, but this one is worth trying for an important occasion. It is similar to the Sauce Hollandaise,* only stronger in flavor. Use it for meat fondue, steaks, broiled and baked fish, and all egg dishes.

¼ cup wine vinegar
¼ cup dry white wine
1 tablespoon green onions, finely chopped
2 teaspoons tarragon
4 egg yolks
Pinch pepper
1 ½ sticks butter, melted
Salt to taste
1 ½ tablespoons green parsley, finely chopped

Bring the vinegar and wine to a boil with onions and tarragon. Cook at medium heat until it reduces to 2 tablespoons.

Beat the egg yolks for 3 minutes; add wine mixture and pepper. Beat for 1 more minute. Pour in the melted butter in a thin stream, still beating. Season with salt and parsley. Beat until thick.

SERVES 6.

- -

Sauce Bordelaise from France

This delicious marrow sauce is used primarily for steaks and roasts. It also makes an excellent sauce for meat fondue. Ask the butcher to split the bones for you.

Marrow from two large beef bones
⅓ cup small onions, finely chopped
4 tablespoons butter
1 cup beef bouillon from cubes
1 ⅓ cups dry red wine
2 teaspoons cornstarch
2 teaspoons water
Salt and pepper to taste
⅓ cup green parsley, finely chopped

Dig the marrow out of the bones. Fry the onions in the butter on medium heat until transparent. Heat the bouillon with the wine, add to the onions. Cook at high heat until reduced to half of the volume. Stir in the cornstarch combined with water, marrow, and parsley. Season.

SERVES 6.

String Beans Polonaise

Each nationality has its own favorite way of serving vegetables. The English boil them thoroughly and drain. The Italians prefer to drown them in a tomato sauce and bake. The French like to hide them in a sauce Hollandaise or sauce Béchamel. Poles too have their own way; they cook many of their vegetables in a little water to which they have added a bit of sugar, and then, on a serving platter, they sprinkle them generously with bread crumbs fried in butter.

In Poland, Brussels sprouts, whole leaves of savoy cabbage, young whole carrots, cauliflower, and asparagus are served the same way as these string beans.

2 pounds string beans
1 cup water, boiling
Salt to taste
½ teaspoon sugar
4 tablespoons butter
4 tablespoons toasted bread crumbs

Cut off the ends, rinse, and cook the beans in boiling water with salt and sugar for 15 minutes. Drain.

Melt the butter on medium heat; add the bread crumbs. Fry, stirring until golden brown. Pour over the vegetable.

SERVES 8.

. . . .

Carrots with Peas from Vienna

The Viennese cuisine greatly influenced the tables of the Central European nations. After the annexation of neighboring territories by the Austrian monarchy, Hungarian, Polish, and Czech leaders and nobility often visited Vienna, taking back home the culinary novelties they had enjoyed there.

2 packages frozen carrots, diced
1 package frozen green peas
2 tablespoons butter
2 tablespoons instantized flour
¾ cup chicken bouillon from cube
½ teaspoon sugar
2 teaspoons green parsley, chopped
Salt to taste

Cook the carrots and the peas. Combine.

Melt the butter at medium heat and add the flour; fry,

stirring until pale golden. Stir in the bouillon. Add the vegetables. Season with sugar and salt. Sprinkle with parsley.

SERVES 8.

. . . .

Spinach Creamed the Continental Way

There is no food in the world worse than overcooked, soggy, and tasteless vegetables. This modern, quick way of preparing spinach prevents the loss of vitamins and preserves its beautiful green color.

Spinach is served in Europe for dinner with roasts and fried meats just as it is here, but it also makes a good luncheon dish in combination with fried eggs.

2 pounds spinach
3 cups boiling water
Salt
2 tablespoons flour
¾ cup milk
Dash garlic powder

Rinse the spinach thoroughly in a colander. Place in a saucepan, add boiling water, and salt. Bring to a boil. Cook at high heat for 3 minutes. Drain. Place in a blender. Add flour and milk. Turn blender to grate for 2 seconds. Return to the saucepan, bring to a boil, cook for 2 minutes. Season with salt and garlic.

SERVES 6.

. . . .

. . . .

Spinach the Belgian Way

Belgian cooking is similar to French, but has been also influenced by the German and Dutch. Some regional Belgian dishes are original and worth trying.

Spinach, so often rejected by the young, may become a tastier and more appealing vegetable when served in a more inspired way.

4 cups spinach
1½ tablespoons grated onion
½ cup sour cream
3 tablespoons prepared horseradish
¾ teaspoon salt
Dash of pepper

Cook spinach in a large kettle of boiling salted water for 3 minutes. Drain in a colander. Add all the other ingredients and grate coarsely in the blender.

Reheat, but do not boil.

SERVES 6.

. . . .

Spinach Soufflé from Grenoble

This is the favorite dish of my childhood. Our cook was a real wizard with soufflés, and mother ordered one whenever she wanted a really elegant meal. In our home spinach soufflé was usually served with roast veal or Veal Schnitzels.* Tomato or red cabbage salad made a colorful contrast.

Remember that soufflés fall fast after being taken out of

the oven, so make sure that everybody is at the table in time.

2 tablespoons green onions, chopped fine
3 tablespoons butter
¾ cup finely chopped spinach, cooked, and well drained
3 tablespoons instantized flour
1 cup milk
4 egg yolks
5 egg whites
½ teaspoon salt
¼ teaspoon cream of tartar
⅓ cup grated Parmesan cheese

Sauté the onions in 1 tablespoon butter, add spinach and fry for a few minutes, stirring.

Prepare the usual white sauce from the remaining butter, flour, and milk. Remove from the heat, add the egg yolks one by one, stirring briskly. Combine with the spinach.

Whip the egg whites with salt until stiff, add cream of tartar, beat some more. Fold a few tablespoons of egg whites into the spinach mixture. Stir in half of the cheese. Fold in the remaining egg whites. Turn into a deep 1½-quart baking dish which has been buttered and sprinkled with Parmesan cheese. Sprinkle with the remaining cheese. Bake at 375° oven for 35 minutes.

SERVES 5.

Purée of Green Peas from Vienna

Viennese gourmets considered cooking an art no less important than painting, sculpture, or music. They incorporated into their cuisine many elaborate French dishes, but also aromatic stews of Hungarian herdsmen, original soups from the Carpathian mountains, the Czechs' love for pork and geese, Polish sweet fantasies, and exotic Turkish spices. It all became a very special Viennese culinary cocktail.

Purée of peas is the perfect vegetable to serve with roast beef.

3 packages frozen young green peas
2 tablespoons butter, melted
Salt to taste
¼ teaspoon pepper
1 egg yolk
⅓ cup coffee cream
1 tablespoon green parsley, finely chopped

Cook the peas and drain. Press through a strainer. Mix with butter; season with salt and pepper. Bring to a boil. Stir in the egg yolk mixed with cream. Transfer to a serving dish. Sprinkle with parsley.

SERVES 8.

. . . .

Finnish Christmas Rutabaga

As in many European countries, Finland's most festive celebration takes place on Christmas Eve. First, of course, there is a smorgasbord of crackers and assorted cheeses, several kinds of canned fish, and small casseroles. But this is merely a prelude to the real thing, which starts with a fish dish, followed by baked ham with rutabaga pudding. Next comes rice cooked with milk, sugar, and cinnamon. As a finale, there are prune tarts and lots of very strong coffee.

Rutabaga is called Swede in Finland. A turnip with a very large yellowish root, it can be purchased in larger food markets.

Serve with baked ham and roast meats.

8 cups rutabaga, peeled, diced
Salt
½ cup bread crumbs
¼ cup cream
2 eggs, slightly beaten
½ teaspoon nutmeg
1 tablespoon butter

Boil rutabaga in a little salted water for 20 minutes. Chop rutabaga with its water in a blender. Soak half of the bread crumbs in cream. Mix with the eggs and nutmeg. Combine with rutabaga. Pour into a well-buttered baking dish. Sprinkle with the remaining bread crumbs; dot with butter. Bake in a 375° oven for 1 hour.

SERVES 10.

Hungarian Baked Asparagus

Hungarians love to dance and to sing to the romantic music of a violin, sentimental at times or gay, reflecting a pleasure in good company and good food. And for Hungarians good food is more or less associated with sour cream. Serve this rich vegetable dish with roasts, broiled or fried fish, and poultry—meats prepared without gravy. Serve with tomato salad for color.

3 packages frozen asparagus
Salt to taste
½ teaspoon sugar
¼ pound butter
½ cup bread crumbs
1 pint sour cream
1 teaspoon paprika

Cook asparagus in a little water with salt and sugar. Drain.

Fry bread crumbs with butter until golden.

Spread half of sour cream on the buttered bottom of a 9-inch-square baking dish. Sprinkle with half of bread crumbs. Arrange the asparagus in even rows. Cover with remaining sour cream. Sprinkle with remaining bread crumbs and paprika. Place under broiler for about 10 minutes or until golden.

SERVES 8.

Russian Red Beets

It is hard to believe that 50 years ago flying fresh greens from the South was only science fiction. People had to be satisfied with root vegetables and cabbage for most of the year, and they used their imagination to make them as palatable as possible. This is a traditional recipe which tastes as good today as it tasted in times when people didn't have many fresh vegetables during the winter. Serve with roasts and baked potatoes.

Two 1-pound cans red beets, drained, finely chopped
½ cup beet juice
2 tablespoons instant flour
Salt
1 teaspoon sugar
1 tablespoon lemon juice
¾ cup sour cream

Combine the beets with beet juice mixed with flour. Bring to a boil. Add salt, sugar, lemon juice, and sour cream. Heat, but do not boil.

SERVES 6.

. . . .

Hungarian Cabbage

There is always a feast at Hungarian weddings. Innumerable dishes are prepared a day or several days ahead in the big family kitchen, or if it is in the country, over open fireplaces. The wedding feast often starts with fruit plucked from an arrangement called the "Tree of Life." Then come fish, soup, poultry, pork, and always several kinds of vegetables. The celebrations take place from afternoon until late at night. In better times they lasted for several days.

Serve cabbage with pork roast, pork chops, or lamb chops.

3-pound head cabbage, thickly sliced
Salt
1 tablespoon caraway seeds
1 large onion, finely chopped
2 tablespoons bacon drippings
2 tablespoons instant flour
1 tablespoon green parsley, chopped

Cook the cabbage in a little salted water with caraway seeds for 20 minutes.

Fry the onions in hot drippings until golden. Stir in the flour and fry for a few minutes. Dilute with the liquid from the cabbage. Add the cabbage, bring to a boil, stirring. Transfer to a serving platter; sprinkle with parsley.

SERVES 8.

Red Cabbage from Germany

I spent my honeymoon in an ex-prisoner-of-war camp in a small town in the vicinity of Hamburg. We were provided with American Red Cross food parcels which made a sufficient, but monotonous diet. One day my husband had a wonderful idea. He made a deal with a local butcher. In exchange for our American goodies—coffee, cocoa, chocolate, spices, and canned and dried fruits, so scarce in Germany right after the war—each day his wife would give us a home-cooked dinner containing fresh meats and vegetables. One of my favorites among her specialties was this delicious red cabbage.

4 slices bacon, chopped
1 medium onion, chopped
3-pound head red cabbage, coarsely shredded
1 cup water
2 apples, peeled, sliced
1 tablespoon sugar
Salt to taste
¾ teaspoon caraway seeds
⅓ cup lemon juice

Fry the bacon with the onions in a saucepan until pale golden. Add the cabbage and water. Cook at low heat for 30 minutes. Add the apples, sugar, salt, and caraway seed. Cook uncovered over high heat, stirring, until all water evaporates. Combine with lemon juice and serve.

SERVES 6.

. . . .

Ukrainian Sauerkraut with Mushrooms

Ukrainians often use wild mushrooms in their cooking. Their tempting taste is much stronger than that of the mushrooms we find in our supermarkets. In Eastern and Central Europe, wild mushrooms are gathered in season, threaded on strings, and dried out for use in winter months. Dried European mushrooms are sold in many gourmet shops in America. If necessary, you may substitute fresh ones for them in this recipe, using 4 ounces for each ounce of dried ones. Fresh mushrooms should be quartered, sautéed in hot fat, and then added to stewed sauerkraut.

1 ounce dried mushrooms, washed
½ cup water
2 pounds sauerkraut, drained and rinsed
2 onions, sliced
3 tablespoons bacon drippings
3 tablespoons flour
1 teaspoon sugar
Salt to taste

Soak the mushrooms in water for 2 hours, slice, save the water.

Place sauerkraut in a small amount of boiling water. Add mushrooms and their water, carefully leaving the sand that gathered on the bottom of the cup. Bring to a boil. Simmer covered for 30 minutes. Refrigerate overnight.

Reheat sauerkraut. Fry the onions in hot fat until golden. Add flour, fry some more, stirring. Dilute with the liquid from sauerkraut. Add to sauerkraut and bring to a boil. Season.

SERVES 8.

· ·

Noodles with Cabbage from the Hungarian Countryside

Culinary habits of every nation change with the times. Some dishes are quickly forgotten, others take their place. Certain favorites remain for many centuries.

Noodles with cabbage is one of those old-timers that has never lost popularity in Hungary. Although of peasant origin, the dish was served in country manors even before the Habsburgs established their rule in the country. Today it is also favored in the cities, especially to complement a pork roast or pork chops.

Serve with tomato salad for a colorful contrast.

1 medium head of cabbage, grated coarsely
2 teaspoons salt
½ pound egg noodles
2 tablespoons bacon drippings or shortening
¼ teaspoon pepper

Mix the cabbage with salt; let stand at room temperature for ½ hour. Squeeze the juice out.

Cook the noodles and drain.

Fry the cabbage in hot drippings on medium heat for 45 minutes, stirring occasionally. Season with pepper and salt. Mix with the noodles, heat, and serve.

SERVES 6.

. . . .

Baked Potatoes from Finland

Finland is best known in America for its sauna and its music. But Finnish cooking is also worth investigating. It may provide a very pleasant surprise, a whole new field of original national dishes of astonishing variety. These potatoes are easy to prepare and always tasty.

6 medium potatoes, peeled
¼ pound margarine, melted
½ cup bread crumbs
½ teaspoon salt

Cut each potato ¾ way through into thick slices. Dip each potato in melted margarine and then in bread crumbs. Bake in 400°F. oven for 1 hour.

SERVES 6.

. . . .

Swedish Spring Potatoes

I have never seen on this continent potatoes as young as those served in Europe in the spring, no more than the size of marbles or walnuts. They have an incomparably delicate taste. It is possible, however, to find here in May larger, egg-sized young potatoes, which are the next best thing. Never peel baby potatoes; they should only be scrubbed vigorously with a sharp brush.

2 pounds young potatoes, well scrubbed
¼ pound butter, melted
3 tablespoons dill weed, fresh or dry

Cook young potatoes in boiling lightly salted water for 15 minutes. Drain. Transfer onto a warmed serving platter. Pour over butter; sprinkle with dill.

SERVES 6.

- -

Bohemian Boiled Potatoes

In Central and Northern Europe, potatoes are served boiled more often than in any other way. This does not mean, however, that they are just boiled and served. They are seasoned in different ways to make a delicious dish. Sometimes they are sprinkled with bread crumbs fried in butter, or with caraway seeds and pork-roast drippings. Bohemians like bacon with their potatoes, especially during the winter when the potatoes are starchy and crumbly.

2 pounds medium potatoes, peeled, cut into three pieces
1 bay leaf
6 slices bacon, chopped
2 tablespoons green parsley, chopped

Boil potatoes in salted water with the bay leaf for 20 minutes. Drain. Let stand uncovered for a few minutes to let the steam out. Cover and shake vigorously until potatoes fall apart. Transfer onto a warmed serving dish. Sprinkle with fried bacon, drippings, and parsley.

SERVES 6.

- -

Swedish Apple Squares

Our American apple pie has its roots in northern Europe.
Sweden can offer its own delectably different variation. Serve
it warm or at room temperature on cool autumn evenings
when the air is crisp.

DOUGH
2 cups flour, sifted
¼ pound margarine
½ teaspoon baking powder
⅔ cup confectioners' sugar
1 teaspoon vanilla extract
2 large egg yolks

STUFFING
3 pounds tart apples
2 tablespoons sugar
Juice ½ lemon
3 tablespoons bread crumbs
⅔ cup walnuts, chopped

TOPPING
1 cup coffee cream
3 eggs, separated
⅔ cup confectioners' sugar
1 tablespoon cornstarch
1 teaspoon grated lemon rind

TO MAKE DOUGH: Sift the flour before measuring. Cut
the margarine into the flour with a knife and rub in with
fingertips, add the sugar and the baking powder; mix. Add

vanilla mixed with egg yolks. Knead the dough. Spread the dough on the bottom and the sides of a 9" × 12" pan, buttered and sprinkled with flour.

TO MAKE STUFFING: Peel and core the apples; cut each into 8 wedges. Sprinkle with sugar and lemon juice. Sprinkle bread crumbs over the dough in the pan. Arrange the apples on top of the bread crumbs. Dot with walnuts.

TO MAKE TOPPING: Mix the cream with egg yolks, sugar, corn starch, and lemon rind in a blender. Beat the egg whites in a mixer until stiff. Add to the cream mixture, mix slightly with a spatula, pour evenly over the apples.

TO BAKE: Place in a 350° oven for 1 hour. Remove, cool. Cut while still in the pan.

YIELDS 24 SQUARES.

Hungarian Apple Strudel

I often suspect that Austria's two-century-long battle to take Hungary from the Turks was over the strudel. After all, people have made wars for more trivial reasons. And we have to admit that the many varieties of Hungarian strudel are worth fighting for. Anyhow, the Austrians won, and strudel has become as much part of Vienna as it is of Budapest.

Unfortunately we seldom have time or patience nowadays for the tedious kneading and stretching of strudel dough until it is paper thin. The next best is to use a ready-made strudel dough from a supermarket, then stuffing and baking it at home. Even using the prepared dough, it is still very, very good.

Thin sheets of strudel dough
¼ cup butter, melted
½ cup toasted bread crumbs
3 pounds apples, peeled, thinly sliced
⅓ cup raisins
¾ cup sugar
1 teaspoon cinnamon

Spread the dough with butter, sprinkle with bread crumbs. Combine apples with raisins, sugar, and cinnamon. Spread over the dough and roll up. Place on a buttered cookie sheet. Bake in 450° oven for 40 minutes. Serve warm.

SERVES 10.

Viennese Apple Strudel

Strudel dough comes in boxes that are kept refrigerated in the dairy sections of many supermarkets. It can also be purchased in gourmet shops specializing in imported cheeses and European foods. You will probably find it right among the cheeses.

Strudel dough dries rapidly. Do not open the package before the stuffing and everything else is prepared. Work quickly. Once you know where to buy strudel dough, you'll seldom bake anything else because this dough is a terrific labor saver and offers many possibilities for desserts and hors d'oeuvres. Strudels taste best when eaten the same day they are prepared.

Two sheets strudel dough
¼ cup butter, melted
½ cup toasted bread crumbs
1 cup walnuts, chopped
3 tablespoons sugar

STUFFING
5 apples, peeled, thinly sliced
½ cup sugar
⅓ cup raisins

TO MAKE STUFFING: Combine the apples with sugar and raisins.

TO MAKE STRUDEL: Spread a damp dish towel on a table. Unfold strudel dough. Place one sheet on the towel. Brush with half of the butter. Place the second sheet over it. Brush with the remaining butter. Sprinkle with bread crumbs mixed with nuts.

Place the stuffing along the short edge in a 4-inch strip. Roll with the aid of the towel. Place on a buttered cookie sheet. Sprinkle with sugar. Bake in 400° oven for 30 minutes.

SERVES 8.

Almond Strudel from Salzburg

When I was a little girl, one couldn't buy shelled nuts. Beautifully ornate silver nutcrackers were served, and everyone had to help himself. Before making a nut filling for a cake, a cook had to work for a long time to get all the meaty insides from those hard little devils. But I have to admit that our shelled almonds, especially slivered or chopped ones, are often dried out, and don't have much flavor. I like to buy my shelled almonds in protective cans and chop or grind them myself.

This cake tastes best freshly baked. Reheat if you have to serve it next day.

2 sheets of strudel dough
¼ cup butter, melted

STUFFING
6 eggs, separated
⅓ cup sugar
¾ cup almonds, ground
1 teaspoon grated lemon rind
1 tablespoon grated orange rind

TO MAKE STUFFING: Beat the egg yolks with the sugar until creamy. Mix the almonds with lemon and orange rinds. Sprinkle the almonds over the stiffly whipped egg whites. Fold in the egg yolk mixture; mix slightly.

TO MAKE STRUDEL: Spread a large damp cloth napkin on a table. Unfold strudel dough. Place one sheet on the napkin. Brush with half of the butter. Place the second sheet over it. Brush with the remaining butter. Place the stuffing along the short edge in a 4-inch strip. Roll with the aid of the napkin. Place on a buttered cookie sheet and bake in 400° oven for 30 minutes.

SERVES 8.

. . . .

Cheese Strudel from Vienna

Strudel dough may be used for the crust of all kinds of pies and cheese cakes. However, for a true Viennese-style cheese strudel, you have to roll it and bake as a long loaf. Serve with vanilla ice cream. It tastes best when it has just been baked.

Two sheets strudel dough
¼ cup butter, melted
1 teaspoon lemon rind, grated
¼ cup raisins

STUFFING
4 eggs, separated
⅔ cup sugar
2 tablespoons butter
¼ cup sour cream
1 pound farmer cheese

TO MAKE STUFFING: Beat the egg yolks with sugar until creamy, add butter in small bits, constantly beating. Add sour cream and beat some more. Whip egg whites for 10 minutes. Press the cheese through a strainer over the egg whites. Fold in the egg yolk mixture. Mix lightly.

TO MAKE STRUDEL: Spread a damp large cloth napkin on a table. Unfold strudel dough. Place one sheet on the napkin. Brush with half of the butter. Place the second sheet over it. Brush with the remaining butter. Place the stuffing along the short edge in a 4-inch strip. Sprinkle with lemon rind and raisins. Roll with the aid of the napkin. Place on a buttered cookie sheet and bake in 400° oven for 30 minutes.

SERVES 8.

. . . .

Hungarian Walnut Strudel

When fresh berries and plums are just a memory, and a crackling fire keeps us company on cool autumn nights, it is time to use the fresh crop of walnuts for this exquisite strudel.

Two sheets of strudel dough
¼ cup butter, melted
½ cup bread crumbs

STUFFING
12 ounces walnuts, ground
¾ cup sugar
1 teaspoon lemon rind, grated
1 cup coffee cream
½ cup raisins

TO MAKE STUFFING: Boil the walnuts with sugar, lemon rind, and cream, stirring until thickened. Mix with raisins, cool.

TO MAKE STRUDEL: Spread a damp large cloth napkin on a table. Unfold strudel dough. Place one sheet on the napkin. Brush with half of the butter. Place the second sheet over it. Brush with the remaining butter. Sprinkle with bread crumbs. Place the stuffing along the short edge in a 4-inch strip. Roll with the aid of the napkin. Place on a buttered cookie sheet and bake in 400° oven for 30 minutes.

SERVES 8.

Hungarian Poppyseed Strudel

Lots of poppies are grown in Central Europe without any intention of concocting illegal drugs. The seeds are commonly used for stuffing cakes and sprinkling breads.

Two sheets of strudel dough
¼ cup butter, melted
½ cup bread crumbs

STUFFING
½ pound poppyseeds
¾ cup sugar
1 cup milk
3 apples, coarsely grated
¼ cup raisins
1 teaspoon grated lemon rind

TO MAKE STUFFING: Grate the poppyseeds in a blender in 4 portions. Boil with sugar and milk, stirring until slightly thickened. Mix with remaining ingredients, cool.

TO MAKE STRUDEL: Spread a damp large cloth napkin on a table. Unfold strudel dough. Place one sheet on the napkin. Brush with half of the butter. Place the second sheet over it. Brush with the remaining butter. Sprinkle with bread crumbs. Place stuffing along the short edge in a 4-inch strip. Roll with the aid of the napkin. Place on a buttered cookie sheet and bake in 400° oven for 30 minutes.

SERVES 8.

Hungarian Blueberry Strudel

Centuries before people in Vienna learned to make strudel, and added some of their own refined variations, Hungarians were serving it on warm summer evenings, using every possible fruit or berry in season for stuffing.

Two sheets of strudel dough
¼ cup butter, melted
⅔ cup toasted bread crumbs
3 cups blueberries
1 cup sugar

Spread a damp large cloth napkin on a table. Unfold strudel dough. Place one sheet on the napkin. Brush with half of the butter. Place the second sheet over it. Brush with the remaining butter. Sprinkle with bread crumbs. Place blueberries mixed with sugar along the short edge in a 4-inch strip. Roll with the aid of the napkin. Place on a buttered cookie sheet and bake in 400° oven for 30 minutes.

SERVES 8.

- -

Cherry Strudel from Vienna

It doesn't take very long to pit the cherries for this strudel and the result is really delightful. This cake tastes best when freshly baked. Serve warm with vanilla ice cream.

Two sheets strudel dough
¼ cup butter, melted
½ cup toasted bread crumbs
⅔ cup almonds, finely chopped
2 cups cherries, pitted
½ cup sugar
½ cup sour cream
3 tablespoons sugar

Spread a damp large cloth napkin on a table. Unfold strudel dough. Place one sheet on the napkin. Brush with half of the butter. Place the second sheet over it. Brush with the remaining butter. Sprinkle with bread crumbs mixed with almonds.

Distribute the cherries evenly along the short edge covering ⅓ of the dough. Sprinkle with sugar. With the aid of the napkin, roll strudel like a jelly roll. Place on a buttered cookie sheet. Spread with sour cream. Sprinkle with sugar. Bake in 400° oven for 30 minutes.

SERVES 8.

. . . .

French Cherry Pie

It takes only 15 minutes to prepare this pie. The French don't even pit the cherries. They believe that the stones add a special desirable flavor to the pie. I prefer to use pitted cherries, as I would hate to see one of my guests break a tooth on my dessert.

12 ounces cherries
1 cup flour
⅔ cup sugar
3 eggs
1 cup milk
Pinch of salt

Place the cherries evenly in a well buttered 8-inch pie pan. Beat the remaining ingredients in the mixer for 5 minutes. Pour over the cherries. Bake in a 350° oven for 45 minutes.

SERVES 4.

. . . .

Apple Pie from Switzerland

The idea of a fruit pie covered with custard is very Swiss. This is a favorite Sunday treat of Swiss families, not too sweet, not too heavy, but healthful and nourishing.

DOUGH
1 ¾ cups flour
Pinch salt
2 tablespoons confectioners' sugar
¼ pound and 1 tablespoon sweet butter
1 egg

TOPPING
3–4 apples, peeled, sliced
½ cup milk
1 ½ teaspoons cornstarch
2 tablespoons sugar
½ cup whipping cream
1 egg

TO MAKE DOUGH: Combine the flour with salt and sugar. Cut the butter into the flour and crumble with your fingertips. Add the egg and knead the dough quickly. Refrigerate for ½ hour.

Roll out and transfer in pieces to cover the bottom and the sides of a 9–10 inch pie pan. Pierce in a few places with a fork.

TO MAKE TOPPING: Arrange the apples on the pie crust. Combine the ingredients of the custard in a blender. Pour over the apples.

Bake for 40 minutes in 350° oven. Serve warm.

SERVES 5–6.

Strawberry Pie from Switzerland

Strawberries are wonderfully aromatic and tasty in the early summer. Take full advantage of their season, serving strawberries in as many different ways as possible. The crust of this pie may be baked well in advance and kept frozen. Make two or three at a time and have a ready supply when needed.

CRUST
Prepare the dough as for
*Apple Pie from Switzerland**

TOPPING
½ cup currant jelly
1½ pints medium strawberries, washed and hulled
½ cup whipping cream

TO MAKE CRUST: Roll out the dough and transfer in pieces to cover the bottom and the sides of a 9″-to-10″-inch pie pan. Cover with foil and spread ½ cup of uncooked rice on the foil. Place in a 400° oven for 10 minutes. Reduce the heat to 350°. Bake another 10 minutes. Take out, remove the rice and the foil. Pierce the crust in a few places with a fork. Bake another 10 minutes. Cool.

TO MAKE TOPPING: Warm the jelly to melt. Arrange strawberries on the crust. Brush with jelly. Refrigerate.

Whip the cream and decorate the pie with it or serve in a separate dish.

SERVES 5–6.

. . . .

Dutch Strawberry Cake

A thousand years ago fiercely independent tribesmen lived among the dunes and marshes of the Dutch coast. Subjected to terrible weather and even more terrible neighbors —the Vikings—they held onto this dismal country and started building against the frequent onslaughts of the sea. Thus started the Dutch tradition of building up their low and vulnerable land and cherishing each foot of it for the sweat and blood it took to make and keep it. There isn't a single piece of uncultivated land in the Netherlands. Everything must be used in a productive way. Organic wastes flower in the tulips and smile in the redness and sweetness of Dutch strawberries.

BATTER
3 large eggs
1 cup sugar
⅓ cup coffee cream
1½ cups instant flour
2 teaspoons baking powder

TOPPING
1 quart fresh strawberries, sliced
⅔ cup confectioners' sugar
1 cup whipping cream
3 tablespoons confectioners' sugar
1 teaspoon vanilla extract

TO MAKE BATTER: Beat the egg yolks for 3 minutes. Add sugar in small portions, still beating. Beat 5 more minutes. Add the cream, flour, and baking powder, and beat 5 more minutes. Whip egg whites until stiff. Add 2 tablespoons of the egg whites to the batter, mix. Add the rest of the egg whites, mix lightly.

Spread the batter in two 9-inch round cake pans that have been buttered and wax-paper-lined. Bake in 350° oven for 20 minutes. Cool slightly. Remove from pans. Cool. Cover and refrigerate overnight.

TO MAKE TOPPING: 1½ hours before serving, sprinkle strawberries with sugar. Let stand for 20 minutes. Whip the cream, add sugar and vanilla.

TO MAKE THE CAKE: Sprinkle 1 cake layer with half of the strawberry juice. Cover with ⅔ of the strawberries; spread with half of the whipped cream. Top with the second cake layer. Sprinkle with remaining strawberry juice. Spread with remaining cream. Garnish with the rest of strawberries. Refrigerate for 1 hour.

SERVES 12.

Swiss Chocolate Cake

This is the very first cake I ever baked. My Swiss girl friend from our quite international School of Journalism in Paris gave me her mother's recipe. I was in tears after taking it out of the oven. It was so heavy! I was sure it was a big flop. But I changed my opinion after tasting it. It was really great. This rich cake is supposed to look that way.

BATTER
6 ounces baking chocolate
2 tablespoons water
6 ounces sweet butter
4 eggs, separated
¾ cup sugar
1 ⅔ cups sifted flour
1 teaspoon baking soda

FILLING
2 egg yolks
¾ cup confectioners' sugar
¼ pound sweet butter
3 tablespoons instant coffee
3 ½ tablespoons vodka

SPREAD
6 ounces chocolate, melted
2 tablespoons sweet butter

TO MAKE BATTER: Melt chocolate with water and butter on low heat. Beat the egg yolks with sugar for 7 minutes. Combine with the chocolate mixture. Fold in alternately in small portions whipped egg whites and flour mixed with soda. Mix lightly. Bake in a 9-inch spring pan, which has been buttered and sprinkled with bread crumbs, in 375° oven for 45 minutes. Cool. Cut into 2 layers.

TO MAKE FILLING: Beat the egg yolks with sugar for 5 minutes. Add the butter in small bits while still beating. Beat 5 more minutes. Add coffee dissolved in vodka, beat some more. Spread between the layers of the cake.

TO MAKE SPREAD: Melt the chocolate with butter on low heat; do not let it boil or dry out. Spread over the top and the sides of the cake. Refrigerate.

SERVES 14.

· ·

French Easter Cake

Among the great variety of cakes served in France at Easter, this one is the most popular. It consists of one of the basic French pastries, called *Pâte Genoise,* an exquisite recipe that doesn't require baking powder and will not fall. The trick lies in extensive beating of the eggs with sugar, and mixing the other ingredients only very lightly.

Many kinds of fillings and icings may be used with this pastry. The favorite French butter spread for this Easter Cake is flavored with ⅓ cup of powdered praline—almonds roasted in boiling sugar. This is difficult to duplicate in America as praline may be found here only in special gourmet shops, or you have to go to all the trouble of making it yourself. French housewives buy powdered or chopped praline at the grocer's and use it for many desserts. Similar in taste, the next best, is the creamy butter spread flavored with caramel and toasted almonds offered here.

Bake this Easter Cake a day ahead of your party.

BATTER
4 eggs
⅔ cup sugar
1 teaspoon almond extract
1 cup flour, sifted
4 tablespoons margarine, melted

SPREAD
1 tablespoon sugar
3 tablespoons table cream
2 egg yolks
⅔ cup confectioners' sugar
¼ pound sweet butter, soft
⅓ cup toasted almonds, ground

DECORATION
2 tablespoons toasted almonds, chopped
1 chocolate egg

TO MAKE THE BATTER: Beat the eggs with sugar for 10 minutes. Add almond extract, beat some more. Sift the flour over the eggs. Mix very lightly. Slowly add lukewarm margarine, mixing lightly with a spatula.

Bake in a well-buttered and floured 9-inch-square pan in a 350° oven for 30 minutes. Then turn off the heat, leave the oven door half open, and let the cake stand for 5 minutes.

TO MAKE THE SPREAD: Brown sugar in a small pan. Add the cream and stir till dissolved. Beat the egg yolks with sugar for 5 minutes. Add the butter in small bits, beating constantly. Add the cream and beat some more. Combine with ground almonds.

Cut the cake into two layers. Spread the first layer; cover; spread the top and sides of the cake. Sprinkle with chopped almonds. Place the chocolate egg in the center. Refrigerate.

SERVES 12.

Russian Cheese Cake

Russia has changed considerably from the time of the revolution, and so has Russian cooking. But the old recipes were carried abroad by the refugees, cherished, and have been handed down from mother to daughter. Here is a delicious cheese cake from the old Russian country manors. The original recipe calls for rose-petal jam, which is difficult to find in the United States. Any pure and thick berry preserves may be substituted.

DOUGH
¾ stick margarine
1 ¾ cups sifted flour
½ cup confectioners' sugar
1 ½ teaspoons baking powder
1 egg
½ teaspoon vanilla extract
3 tablespoons sour cream

FILLING

6 *eggs*
2 ¼ *cups confectioners' sugar*
1 ½ *pounds farmer cheese*
1 ½ *sticks sweet butter, soft*
2 ½ *tablespoons instant flour*
1 ½ *teaspoons vanilla extract*
2 *teaspoons grated lemon peel*
¼ *cup candied orange rind, finely chopped*
⅓ *cup raisins*

SPREAD

6 *ounces berry preserves*
1 *egg white*

TO MAKE DOUGH: Cut the margarine into the flour and rub in with fingertips. Add the sugar, baking powder, and egg mixed with vanilla and sour cream. Knead the dough.

Spread ⅔ of the dough on the bottom of a 9″ × 12″ pan, which has been buttered and sprinkled with bread crumbs. Bake at 375° for 10 minutes. Cool.

TO MAKE FILLING: Beat the eggs with the sugar for 5 minutes. Press the cheese through a strainer; combine with butter, vanilla, and flour. Add to the eggs. Fold in lemon peel, orange rind, and raisins.

Spread the cake with preserves. Spread the cheese filling over it. Form long, thin rolls from the remaining dough, using them to form a diagonal criss-cross crust on the cake. Spread with the egg white. Bake at 375° for 1 hour. Cool. Cut into squares in the pan.

SERVES 20.

. . . .

Italian Cheese Torte

Italians are not just spaghetti and pizza eaters. They originated magnificent desserts, light and tasty, not overly sweet. Here is an Italian torte based on a simple sponge cake. It is a time and effort saver because a sponge cake from your favorite baker may be used with good result. It also can be prepared a day ahead.

BASE
1 plain 8-inch sponge cake

PUNCH
4 tablespoons orange curaçao
½ cup water

FILLING
3 egg yolks
½ cup confectioners' sugar
¾ pound farmer cheese
3 tablespoons butter
½ cup chocolate chips, finely grated in a blender
1 tablespoon orange curaçao

SPREAD
1 cup whipping cream
2 tablespoons confectioners' sugar

Place the cake on a serving platter. Using a long, thin knife cut it carefully into two layers. Sprinkle the first layer with half of the punch.

TO PREPARE THE FILLING: Beat the egg yolks with sugar

for 5 minutes. Add the cheese and the butter in small portions, beating constantly. Beat for 5 more minutes. Spread the first layer of the cake with half of the cheese mixture. Refrigerate for ½ hour.

Add ¾ of the chocolate and 1 tablespoon curaçao to the rest of the cheese mixture. Mix well. Spread over the cake. Cover with the second layer. Sprinkle with the rest of the punch. Cover and refrigerate.

BEFORE SERVING: Whip the cream, then add the sugar to it. Spread over the top and sides of the torte. Sprinkle with the rest of the chocolate.

SERVES 12.

Viennese Sacher Torte

Each European nation takes credit for several dishes that have gained international fame. One with real éclat is Sacher Torte, a masterpiece of Viennese cookery. Its very name makes you think of charming little restaurants hidden in green gardens, of waltzes, crinolines, laughter, and the naughty Viennese maidens.

There are several popular methods for preparing this incomparable pastry. The following one has one important quality—it never fails.

BATTER
10 egg whites
⅛ teaspoon salt
¾ cup sugar
¼ pound sweet butter
6 ounces chocolate
8 egg yolks
1 tablespoon lemon rind, grated
1½ teaspoons cinnamon
¾ teaspoon cloves
1 cup flour, sifted

SPREAD
¾ cup pure apricot jam

ICING
3 egg yolks
⅔ cup confectioners' sugar
½ cup sweet butter, soft
6 ounces chocolate, melted
2 tablespoons vodka

TO MAKE BATTER: Beat the egg whites with salt until stiff. Add sugar by spoonfuls, beating constantly. Beat 4 more minutes.

Melt butter with chocolate on low heat, stirring. Take off the heat, cool a little. Stir in the egg yolks, lemon rind, cinnamon, and cloves.

Add ¼ of the egg whites to the chocolate mixture; mix lightly. Pour this mixture over the remaining egg whites and sift the flour over it. Mix lightly with a spatula until all the flour is moistened. Pour into three round 9-inch cake pans which have been buttered and wax-paper-lined. Bake for 25 minutes in a 350° oven. Then turn off the heat, leave the oven door half open, and let the cake stand for 5 minutes. Cool slightly, remove from the pans, and finish cooling. Spread apricot jam between layers.

TO MAKE ICING: Beat the egg yolks with the sugar for 5 minutes. Add the butter in small bits, beating constantly. Add the chocolate, beat some more. Combine with vodka. Spread on the top and the sides of the torte.

SERVES 16.

. . . .

Coffee Torte from Paris

Did you ever try to decorate your cakes with fresh flowers? It is a great idea. Stick one rose, carnation, or a mum with a very short stem in the middle of this torte and note the effect on your guests.

The French use this type of pastry for many kinds of tortes, changing just the flavor of the spread. The creamy coffee torte is one of their best and most original.

It is practical to cut a ½-inch deep mark on the side of the cake before cutting it into layers for easier matching later.

Bake this torte a day ahead of your party.

BATTER
4 eggs, separated
⅔ cup sugar
1 cup flour, sifted
3 tablespoons margarine, melted

COFFEE SPREAD
3 egg yolks
1 cup confectioners' sugar
¾ cup sweet butter, soft
4 tablespoons water
3 tablespoons instant coffee

TO MAKE THE BATTER: Beat the egg whites until stiff but not dry. Beat the egg yolks with sugar for 7 minutes. Sift the flour over the egg yolks, alternately adding egg whites. Mix very lightly just until all the flour is moistened. Add lukewarm margarine, mix lightly. Transfer immediately into a buttered and floured 9-inch round spring pan. Bake in 325° oven for 40 minutes. Then turn off the heat, leave the oven door half open, and let the cake stand for 5 minutes.

TO MAKE SPREAD: Beat the egg yolks with sugar for 5 minutes. Add the butter in small bits, beating constantly. Add coffee dissolved in water in a small stream, still beating.

Cut the cooled torte into 3 layers. Spread between the layers, on the top, and along the sides of the torte. Cover, refrigerate.

SERVES 12.

French Torte Curaçao

The French flavor their tortes with all kinds of liqueurs, changing the decorations accordingly. A torte with a kirsch spread looks appetizing garnished with candied cherries. Use candied lemon peel for Benedictine spread, green fruits with mint tortes.

This cake may be baked a day ahead, but it is better to spread and decorate it in the afternoon before the party.

*One 9-inch round cake
as for Coffee Torte**

CURAÇAO SPREAD
3 egg yolks
1 cup confectioners' sugar
¾ cup sweet butter, soft
4 tablespoons curaçao

DECORATION
1 can mandarin oranges, drained

TO MAKE SPREAD: Beat the egg yolks with sugar for 5 minutes. Add the butter in small bits, beating constantly. Add curaçao in a thin stream while still beating.

Cut the cooled cake into 3 layers. Spread between the layers, on the top, and along sides of the torte. Garnish with mandarin oranges. Cover; refrigerate.

SERVES 12.

· ·

Grape Torte from Moravia

Czech women are extremely fastidious about the cleanliness of their homes. Once in a Moravian town, the wife of a prosperous businessman invited me to a party. When the day came, it was very rainy, and to my great disappointment the party was postponed. The hostess couldn't stand the idea of her carpeting being soiled by wet feet.

However the party did take place on the next day, and I had a wonderful time. A delicious grape torte was served for dessert.

DOUGH
¼ pound sweet butter
2 tablespoons sugar
3 egg yolks
1 teaspoon vanilla extract
1 ½ cups sifted flour

TOPPING
3 egg whites
Pinch salt
¾ cup confectioners' sugar
½ cup walnuts, ground
1 pound seedless grapes
1 cup whipped cream

TO MAKE DOUGH: Cream the butter with sugar, add egg yolks and vanilla, and beat for 5 minutes. Add the flour and knead the dough. Line a buttered 9-inch spring pan with the dough. Bake in 450° oven for 15 minutes.

TO MAKE TOPPING: Beat the egg whites with salt until stiff, add sugar by spoonfuls, still beating. Beat 5 more minutes. Fold in nuts. Spread over the cake. Arrange grapes on the top. Bake in 325° oven for 55 minutes. Cool. Decorate with whipped cream before serving.

SERVES 12.

· ·

Rum Torte from Vienna

Here I propose that you do a little cheating. You may bake your favorite sponge cake or buy one, using it as a base for the rum torte. But I guarantee that after the treatment you will then give it, not only your friends, but its own mother would never recognize it as a sponge cake.

This is an elegant torte for special occasions. It can be prepared a day ahead and kept under cover in a cool place.

BASE
One 9-inch plain sponge cake

PUNCH
1 cup water
Juice of 1 lemon
⅓ cup sugar
2½ tablespoons rum

APRICOT FILLING
4 ounces pure apricot preserves

ALMOND FILLING
6 ounces almonds, ground
1 cup confectioners' sugar
Grated rind of 1 lemon
½ cup sour cream

ICING
1 cup confectioners' sugar
2½ tablespoons rum

Mix all the ingredients of the punch and stir till dissolved.

Place the cake on a serving platter. With a long thin knife, cut the cake carefully into 3 layers.

Sprinkle the first layer of the cake on the platter with ⅓ of the punch. Spread with apricot preserves. Cover with the second layer; sprinkle with half of the remaining punch.

Mix all the ingredients of the almond filling. Spread over the cake. Cover with the third layer. Sprinkle with the rest of the punch. Mix sugar with rum for the icing. Spread over the top of the torte. Refrigerate several hours. Cut into small pieces.

SERVES 20.

. . . .

Mocha Torte from Warsaw

It is hard to believe, but this torte, worthy of a king's table, gained its fame during World War II. Food was very scarce in Warsaw after Germans occupied the city, and most of the numerous bakeries were closed or sold pastry made of artificial ingredients that was almost impossible to eat. Many families were left without income as husbands and fathers fought abroad or were kept in camps. Out of desperation, a few ladies who had never worked a day in their lives decided to open a little café and sell the only item they knew how to prepare well—elegant tortes. They baked them under most difficult conditions, without any equipment or help, but the result was exquisite, and the little café became famous throughout the city.

Many have since tried without success to duplicate the excellence of their Mocha Torte. After the war, one of these ladies left Poland to join her husband in the United States and she gave me her recipe.

BATTER
8 egg whites
2 ⅓ cups confectioners' sugar
1 tablespoon lemon juice
10 ounces almonds, peeled, ground

COFFEE SPREAD
1 ¼ cups sweet butter
2 egg yolks
⅓ cup confectioners' sugar
2 tablespoons instant coffee
3 tablespoons vodka

TO MAKE BATTER: Beat egg whites until very stiff. Add sugar by spoonfuls, beating for 3 minutes more. Add lemon juice, beat 5 more minutes. Add the almonds, mix slightly.

Spread the batter in 2 brown-paper-lined 9-inch round cake pans. Bake in a 275° oven for 1½ hours. Then turn off the heat and leave the cake in the oven overnight.

TO MAKE SPREAD: Cream the butter in a mixer with egg yolks and sugar. Dissolve the instant coffee in the vodka. Add gradually to the butter mixture. Spread the icing between the layers and on the top of the torte. Refrigerate for a few hours.

To serve, cut with a thin sharp knife, using sawing motions.

SERVES 10.

. . . .

Poppyseed Torte from Vienna

Viennese pastry is Austria's chief contribution to international gastronomy. This torte probably had its origin in Poland since poppyseed is such a popular ingredient in Polish pastry fillings. At the time when the Austrian Empire incorporated large territories of other nations, it borrowed freely from the culinary traditions of the conquered peoples and gave them back to the world in a more refined rendition.

BATTER

1 pound poppyseed
2 ⅓ cups sugar
½ pound sweet butter
8 eggs, separated
2 tablespoons honey
½ teaspoon almond extract
1 tablespoon rum
¼ cup candied orange rind, finely chopped
½ cup raisins
⅓ cup almonds, ground

PUNCH

2 cups water
1 cup sugar
⅓ cup rum

FILLING

½ pound sweet butter, soft
2 cups confectioners' sugar
3 egg yolks
3 tablespoons strong coffee

TO MAKE BATTER: Cover poppyseeds with boiling water in a saucepan. Let stand overnight. Drain, grind 3 times with 1 cup sugar. Cream the butter with the remaining sugar. Add egg yolks, honey, almond extract, and rum. Beat for 5 minutes. Combine with poppyseeds. Fold in whipped egg whites and combined dry ingredients alternately in small portions. Mix lightly. Bake in 9-inch spring pan, which has been buttered and sprinkled with bread crumbs, for 50 minutes in 375° oven. Then turn off the heat and let the torte stand in the oven with the door half opened for 10 minutes. Cool. Cut into three layers.

TO MAKE PUNCH: Cook water and sugar into light syrup. Add rum.

TO MAKE FILLING: Cream the butter with sugar for 7 minutes, adding egg yolks one by one and coffee by spoonfuls.

TO MAKE THE TORTE: Sprinkle the bottom layer of the torte with ⅓ of the punch. Spread with half of the filling. Cover with the middle layer. Sprinkle with half of the remaining punch. Spread with remaining filling. Cover with top layer, sprinkle with the rest of the punch. Refrigerate. Cut into small pieces.

SERVES 20.

Walnut Torte from Poland

The large waves of Polish immigrants that came to America at the beginning of this century consisted mostly of the poorest peasants. They brought with them the memory of just a few dishes their families could afford on holidays, and they did not know much about the cookery of Polish country manors or city dwellers. Traditional Polish cuisine especially abounds in delicious pastries.

The following elegant recipe is easy to make. You just need a light hand in mixing egg whites with the batter. Do not worry if the torte falls down a little after you take it out of the oven. It is supposed to do so.

BATTER
8 eggs, separated
2 cups confectioners' sugar
Juice of 1 lemon
1 teaspoon vanilla extract
3 tablespoons bread crumbs
½ pound walnuts, ground

CHOCOLATE SPREAD
3 egg yolks
⅔ cup confectioners' sugar
½ pound sweet butter, soft
6 ounces chocolate, melted
1 tablespoon instant coffee
2 tablespoons vodka

TO MAKE BATTER: Beat the egg yolks with sugar for 5 minutes, add the lemon juice. Beat the egg whites until stiff. Add vanilla.

Fold into the egg yolk mixture alternately in small portions without mixing: egg whites, bread crumbs, and walnuts. Mix all together lightly.

Spread the batter in 3 buttered and wax-paper-lined 9-inch round cake pans. Bake in 350° oven for 40 minutes. Remove from the oven; cool in the pan.

TO MAKE SPREAD: Beat the egg yolks with the sugar for 5 minutes. Add the butter in small bits, beating constantly. Add the chocolate, beat some more. Combine with coffee dissolved in vodka.

Spread between the layers, on the top, and on the sides of the torte. Decorate with candied fruits.

SERVES 16.

· ·

Black Torte from Fürstenau

I'm the girl who grew her own wedding bouquet and baked her own wedding cake. I was married shortly after World War II in a small German town where you had to make yourself whatever you needed. I must admit that my German landlady gave me the recipe and sound directions for baking the torte.

BATTER
5 egg yolks
2 cups confectioners' sugar
8 ounces black walnuts, ground
½ teaspoon baking soda
¾ cup sifted instant flour
4 egg whites

SPREAD
1 pint whipping cream
3 tablespoons very strong coffee
4 ounces instant chocolate powder
½ cup black walnuts, chopped

TO MAKE BATTER: Beat the egg yolks for 3 minutes, add sugar in small portions, still beating, and beat 5 more minutes. Whip the egg whites until stiff. Combine walnuts with soda and flour. Fold in the dry ingredients alternately with whipped egg whites in small portions, mix everything lightly. Bake in a 9-inch spring pan, which has been buttered and sprinkled with bread crumbs, for 45 minutes in a 350° oven. Cool. Cut into two layers.

TO MAKE SPREAD: Whip the cream, add coffee, and mix. Add chocolate. Spread between the layers, on the top, and sides of the torte. Sprinkle the top with nuts. Refrigerate for 10 minutes and serve.

SERVES 16.

Almond Torte from Corsica

I once spent a wonderful Easter in Corsica. The whole island looked like a basket full of flowers, pink from almond trees, yellow and white from daffodils, blue from blossoms hanging on fences. After walking all day in the meadows and sitting on picturesque rocks by the seashore, we had a wonderful dinner at a local inn. A specialty of southern France—almond torte—was served for dessert.

BATTER
8 eggs
2 cups confectioners' sugar
½ teaspoon almond extract
⅔ cup instant chocolate powder
3 tablespoons bread crumbs
1 ⅔ cups almonds, ground in skins

SPREAD
½ pound sweet butter
1 ¼ cups sugar
2 eggs
1 tablespoon rum
1 cup toasted almonds, ground

TO MAKE BATTER: Beat the egg yolks with sugar and almond extract for 7 minutes. Whip the egg whites until stiff. Add the egg whites and combined dry ingredients to the egg yolks alternately in small portions. Mix lightly.

Bake in 375° oven for 1 hour in a 9-inch spring pan which has been buttered and sprinkled with bread crumbs. Then turn off the heat and let the torte stand for 10 minutes with oven door half opened.

TO MAKE SPREAD: Cream the butter with the sugar for 10 minutes. Add the eggs one by one, still beating. Add rum and beat 5 more minutes. Fold in almonds.

Spread over the top and the sides of the torte. Refrigerate. Decorate with fresh flowers before serving.

SERVES 16.

. . . .

French Christmas Log

The tradition of a Christmas log came from pagan times, when winter celebrations took place around a huge burning log brought from the woods especially for this purpose. Nowadays a cake shaped to look like a log often takes the place of a real one.

Traditionally French people eat their Christmas supper after returning from the Midnight Mass. The meal consists of oysters, goose liver pâté, various salads, vegetables, and more often a roast goose than a turkey. Christmas log makes the dessert. Recently the Catholic Church changed the schedule of its services and Christmas Eve supper is often served earlier.

BATTER
4 egg yolks
⅓ cup sugar
3 egg whites
¾ cup flour
2 tablespoons butter, melted

COFFEE SPREAD
Coffee Torte from Paris*

CHOCOLATE SPREAD
3 egg yolks
⅔ cup confectioners' sugar
½ pound sweet butter, soft
6 ounces chocolate, melted

TO MAKE BATTER: Beat the egg yolks with the sugar until light and creamy. Beat the egg whites until they form stiff peaks. Add to the egg yolks in small portions, alternating with the flour. Rapidly and delicately fold together. When almost blended, add lukewarm butter in two parts, mixing gently.

Immediately turn the batter onto a jelly-roll pan which has been buttered and lined with waxed paper, then buttered and sprinkled with flour. Spread the batter evenly a little over ½-inch thick. Bake in a 375° oven for 8–10 minutes until very lightly golden. If baked too long it will break when rolled. Cool.

TO MAKE CHOCOLATE SPREAD: Beat the egg yolks with the sugar for 5 minutes. Add the butter in small bits, beating constantly. Add the chocolate, beat some more.

TO MAKE THE LOG: Spread the cake with Coffee Spread. Roll. Trim both ends. Spread the roll with Chocolate Spread, curving designs imitating a trunk of a tree. Refrigerate.

Remove from refrigerator and let stand at room temperature for 1 hour before serving. Cut in small slices.

SERVES 10.

Cocktail Parties

PREFACE

With our busy lives and lack of household help, it isn't always easy to prepare a whole meal for our friends. The simplest solution is a cocktail party for 12 to 20 people. In a fairly large room or in two rooms, such a gathering can be quite pleasant.

A cocktail party is also a favorite way of entertaining for persons who do not like entertaining. They try to squeeze in as many people as possible to whom they feel they "owe" an invitation, regardless of whether they like them or not. But not many of their guests appreciate this form of social life. And I don't blame them. You stand for two hours and your feet are killing you. You get plenty to drink and almost nothing to eat.

You can't hear a word of what a tipsy gentleman with beer-breath insists on telling you. With a stiff grin on your face and an occasional understanding nod, you try to decide whether you should have your ears checked. Then from a cloud of smoke a stranger emerges and smearing your cheeks red with kisses evaluates your appearance.

"My, my! That dress is looking better on you every year!"

Becoming totally hoarse after a while, you are thankful that your hostess marked 7 to 9 on the invitation, stating clearly that after that time she did not wish to be bothered by anyone. Europeans would be aghast at such an unheard of lack of hospitality, but it has a practical side—it provides a sound excuse to leave.

Fortunately not all cocktail parties are like that. If the hostess is careful not to overcrowd her home and she invites people who might like each other—a cocktail party can be a pleasant, not too demanding get-together.

Whenever drinks are served, some food should be also offered besides nuts and chips, as it is not healthy to drink on an empty stomach. Finger snacks are most convenient to eat standing since they do not require forks, knives, or even plates. When more substantial food is served, small paper plates are often used. Europeans favor canapés. I have selected an assortment of cold and hot canapés for this chapter. They should be prepared and arranged on trays before the guests arrive. Carefully covered with plastic wrap and placed in a room where the heat has been turned off, they will stay fresh for several hours.

During the winter especially it is nice to offer one or two warm snacks at a cocktail party. With a larger number of guests it is practical, less time consuming, and less expensive to prepare a small buffet instead of finger foods. Grog or hot wine also makes a pleasant change on a frosty day.

Unless it is strictly a neighborhood party and your guests

don't have to drive to get home, it is considerate to clear the alcoholic beverages and serve coffee, tea, and some pastry before they leave. Cocktails are also often followed by a late buffet with a variety of dishes. In that case only small snacks are served with drinks.

On the Continent people always drink and eat at the same time. Substantial hors d'oeuvres are served with the liquor.

There is generally more mingling and flirting between sexes at European parties. That doesn't mean that wife-swapping is a common practice in the Old World but keeping company with your own wife at a party was always considered poor taste. After all, if a couple wants only one another's company, they don't have to leave home.

Men in Europe are expected to be chivalrous, not just by keeping alive the old custom of hand kissing, but by paying the ladies compliments and being attentive and generally pleasant. It is considered poor manners for men in Paris, Vienna, and Rome to discuss business or cars with a group of other men, leaving the ladies alone at the opposite side of the room. On the other hand, women can't expect a man to be interested in their chatter of children, housekeeping, and fashions.

There is a great deal of dancing and singing at European parties. Regardless of age, when a group gets together, music is a must, and people start moving. After a few drinks, they invariably begin singing together, first the latest hits, then melodies bringing back nostalgic memories and soldiers' ballads. In tune or not, the singing certainly adds to the fun and life of a party.

European husbands are also not supposed to dance with their wives all evening. Two or three times at most. It is the obligation of every gentleman to dance with his hostess, then once with every lady unless the party is large; after that he dances with the ones he likes. If you don't feel like dancing most of the evening, you don't go to a dancing party in Europe.

Europeans never play games at parties once they have outgrown childhood. If they don't dance and don't sing, they will certainly engage in very animated discussion. With the exception of Scandinavians, all others can heat up to such a degree just in talking that an outsider listening from an adjoining room may think that a fight will start any moment. But it is just enjoyable conversation, without which the party would be considered boring.

Ideas for Cocktail Parties

BEVERAGES
1. *Lemon Vodka from Eastern Europe*
2. *Red Cocktail from the Rhine*
3. *English Grog*
4. *Bavarian Eggnog*
5. *Austrian Hot Wine*
6. *Mazagran from Prague*

COLD SNACKS
7. *French Canapés with Ham and Chicken*
8. *French Canapés with Crab Meat*
9. *French Canapés with Eggs*
10. *French Canapés with Anchovies*
11. *French Canapés with Sardines*
12. *French Canapés with Tomatoes*
13. *French Canapés with Beef*
14. *French Canapés with Cheese*
15. *Italian Canapés with Sausages*
16. *Spanish Canapés with Bananas*
17. *Danish Egg Spread*
18. *French Beef Spread*
19. *French Chicken Liver Pâté*
20. *Italian Pork Pâté*
21. *Steak Tartare—A European Fad*

HOT SNACKS
22. *Toasted Cheese Canapés from Hamburg*
23. *French Grilled Canapés with Wine*
24. *Dill Strudel from Prague*
25. *Meat and Mushroom Strudel from Prague*
26. *Sausage Strudel from Bratislava*
27. *Swedish Meatballs*

. . . .

Lemon Vodka from Eastern Europe

Lemon vodka is liked as much by the Russians as by Poles and Ukrainians. It is simple to prepare and certainly original.

In Europe people drink it in small brandy glasses with hors d'oeuvres. Here it would be more practical to serve on the rocks in whiskey glasses. It has a lovely pale yellow color and a very pleasant aroma.

1 lemon
2 tablespoons sugar
⅘ quart vodka

Cut the rind of the lemon very thin, almost transparent, separating it from all the white, bitter parts.

Add the lemon rind and sugar to vodka. Let stand for a week. Strain.

YIELDS ⅘ QUART.

. . . .

Red Cocktail from the Rhine

Two thousand years ago the Romans planted the first grape vines on the Rhine and Moselle river banks. Cooking in this region shows more spirit and grace than in the northern potato-loving Germany. This cocktail was served to local princes after long days of hunting.

⅓ cup sugar
Juice of ½ lemon
1 cup water
1 bottle red wine (the ⅘ quart size)
2 oranges, sliced

Combine sugar, lemon juice, water, and wine. Keep at room temperature. Place an orange slice and a small ice cube in each glass. Pour the cocktail over them and serve immediately.

YIELDS ABOUT 1¼ QUARTS COCKTAIL.

- -

English Grog

Grog was invented by seamen, and became their favorite drink after a long exposure to chilling weather. It is also a perfect relaxant at the end of a tense day, or a sure way to cheer up a crowd tired after winter sports.

6 cubes sugar
6 slices orange
4½ cups hot tea
6 teaspoons cherry brandy
⅔ cup rum

Warm up 6 tumblers by rinsing with hot water. Place a sugar cube and a slice of orange in each tumbler. Pour ¾ cup of hot tea into each. Add a teaspoon of brandy and 1½ tablespoons rum. Serve immediately.

SERVES 6.

- -

Bavarian Eggnog

On cold winter evenings in the Bavarian Mountains, people gather in taverns and around fireplaces with mugs of a powerful beverage that brings warmth to the oldest and most chilled bones. Try it on the night of a fierce blizzard, and see what wonders it can do for you after a whole day of skiing.

4 egg yolks
1 ½ cups sugar
1 quart beer
½ teaspoon cloves
½ teaspoon cinnamon

Beat the egg yolks with the sugar till thick and creamy. Heat the beer with the spices. Add the beer to the egg mixture in a thin stream, beating all the time.

Heat while slowly stirring, but do not boil. Serve in mugs.

YIELDS 5 CUPS.

. . . .

Austrian Hot Wine

Austrians always thought themselves more refined than the Germans, using their cooking as an example of their superiority. Viennese cooking is certainly lighter, more delicate, and more diversified than that of Berlin. Drinking habits are also different in Austria where people favor wine over beer. In Vienna you can find hundreds of charming little wine taverns.

For centuries Austrian grandmothers believed that hot wine is an excellent remedy when you are coming down with a cold. Modern medicine will probably not agree completely, but certainly hot wine gives you a warm, comfortable feeling after winter sports or a tiring day.

1 quart red table wine
1 cup sugar
½ teaspoon cloves
¾ teaspoon cinnamon

Combine all the ingredients and heat, but do not boil. Serve in tumblers.

YIELDS 4 CUPS.

Mazagran from Prague

In Europe there are many different ways of serving coffee. Generally it is much stronger than American coffee. People drink it for breakfast with plenty of milk or cream, and black in demitasse cups in the afternoon or after dinner. The Turks grind it so finely that it becomes very thick. In Vienna it is often served with ice cream or whipped cream. On summer days, in garden cafés, all of Central Europe favors iced coffee—but a little more spirited than we are used to having in America. Their version is ideal for backyard parties.

½ cup coffee, regular grind
4 cups water
¼ cup sugar
½ cup cognac

Prepare the coffee, add sugar. Refrigerate for several hours. Add the cognac. Serve in tumblers with ice.

SERVES 4.

. . . .

French Canapés with Ham and Chicken

The purpose of serving canapés is to save your guests the trouble of making their own sandwiches. Small canapés are prettier than a large sandwich and may be consumed in a more esthetic way. In European countries, guests are never supposed to make sandwiches at a party. It would be considered poor manners if someone started piling cold cuts, relishes, and salads on his bread. People help themselves, placing the food on their plates and using their forks. They spread bread with butter and eat it separately. Whenever it is not convenient to have a whole buffet, canapés are most welcome.

10 slices stale white sandwich bread
¼ cup butter, melted
¼ pound butter, soft
1 tablespoon mustard
⅛ teaspoon pepper
5 slices ham, quartered
5 slices chicken breast, quartered
10 slices tomatoes, quartered

Cut off the crusts and divide each slice of bread into 4 squares. Brush with melted butter. Place in 450° oven until lightly toasted. Cool.

Combine butter with mustard and pepper. Spread the toast. Cover half of the canapés with ham, half with chicken. Garnish with tomatoes.

YIELDS 40 CANAPÉS.

. . . .

French Canapés with Crab Meat

Making canapés is an opportunity for using original ideas and artistic expression. Canapés should be not only tasty and varied, but also beautiful in shape and color. They must complement one another, and at the same time provide a delightful contrast. Smoked oysters may be substituted for crab meat.

10 slices stale sandwich bread
¼ pound butter
¼ cup Mayonnaise*
2 hard-boiled eggs, chopped very fine
One 4-ounce can crab meat
2 lettuce leaves

Cut off the crusts and divide each slice of bread into 4 squares. Fry in hot butter in small portions until golden. Cool. Combine the Mayonnaise with the eggs. Spread over each piece of bread. Place some crab meat on top. Garnish with bits of lettuce.

YIELDS 40 CANAPÉS.

French Canapés with Eggs

Canapés should be prepared an hour or two before the party, arranged attractively on trays, tightly covered with plastic wrap, and stored in a cold place until serving time.

Buy a long, narrow loaf of the sliced rye bread that is especially manufactured for party snacks.

40 slices party rye bread
¼ cup salad oil
½ cup Mayonnaise*
6 hard-boiled eggs, sliced
40 slices dill pickles

Brush bread slices with oil. Place in 450° oven until very lightly toasted. Cool.

Spread toast with Mayonnaise; cover with egg slices; garnish with pickles.

YIELDS 40 CANAPÉS.

French Canapés with Anchovies

French canapés come in endless variations. Choose two or three for your party, using the same base of toasted bread, but a different colorful decoration. Mix them on one large tray, arranging in rows of each kind, or serve on separate plates.

10 slices stale white sandwich bread
½ cup salad oil
¼ pound butter, soft
24 fillets of anchovy
40 slices hard-boiled eggs
1 tablespoon green parsley, chopped

Cut off the crusts and divide each bread slice into 4 squares. Fry in hot oil in small portions until golden. Cool.

Combine butter with 4 fillets of anchovy cut into small pieces. Pass through a fine sieve. Spread each piece of bread with anchovy butter. Cover with egg slices. Sprinkle with parsley. Place half of an anchovy fillet diagonally on each canapé.

YIELDS 40 CANAPÉS.

· ·

French Canapés with Sardines

Sardines make tasty snacks. It is best to choose larger ones imported from Norway, Portugal, or France, and canned in oil.

10 slices stale white sandwich bread
½ cup salad oil
¼ cup butter, soft
2 cans sardines
10 slices lemon, peeled, cut into quarters

Cut off crusts and cut each slice of bread into 4 squares. Fry in hot oil in small portions until golden. Cool.

Combine butter with sardines; pass through a fine sieve. Spread each piece of bread with sardine butter. Place a fillet of sardine over it. Garnish with lemon.

YIELDS 40 CANAPÉS.

· ·

French Canapés with Tomatoes

In Europe French canapés are also often served before luncheon or dinner, but only if there is no soup on the menu. French chefs believe that hors d'oeuvres and soup at the same meal would spoil the appetite for the main course.

10 slices stale white sandwich bread
½ cup salad oil
¼ cup butter, soft
4 fillets of anchovy, cut into small pieces
20 slices tomatoes, cut into halves
20 black pitted olives, cut into halves

Cut off the crusts and divide each slice of bread into 4 squares. Fry in hot oil in small portions until golden. Cool. Combine butter with anchovies; pass through a fine sieve. Spread each piece of bread with anchovy butter. Cover with tomato slices, garnish with olives.

YIELDS 40 CANAPÉS.

French Canapés with Beef

Serving canapés may be quite economical if you do not buy caviar and imported cheeses. With imagination, all kinds of leftovers may be used successfully.

10 slices stale white sandwich bread
¼ pound butter
2 cups leftover roast or boiled beef, finely chopped
1 hard-boiled egg, finely chopped
1 medium pickle, finely chopped
1 teaspoon green parsley, finely chopped
¼ cup Mayonnaise*
40 small slices canned beets

Cut off the crusts and divide each slice of bread into 4 squares. Fry in hot butter in small portions until golden. Cool.

Combine beef with egg, pickles, parsley, and Mayonnaise. Pile generously on the toast. Garnish with beets.

YIELDS 40 CANAPÉS.

· ·

French Canapés with Cheese

The idea of melted cheese is not new in America, but it tastes different without a hamburger beneath or with a variation on the usual American cheese. Use good Gruyère cheese from one of the leading American producers.

10 slices stale white sandwich bread
¼ pound butter
40 small slices Gruyère cheese
Paprika
Lettuce leaves

Cut off the crusts and divide each slice of bread into 4 squares. Fry in hot butter in small portions until golden. Place a slice of cheese on each toast. Sprinkle with paprika. Place under broiler until the cheese melts. Serve warm or cold on a serving platter garnished with lettuce leaves.

YIELDS 40 CANAPÉS.

· ·

Italian Canapés with Sausages

French canapés caught on in Europe as elegant, decorative, and convenient party treats. Each nation, of course, added its special foods to thin slices of bread, producing new varieties. Here is an Italian version.

10 slices stale white sandwich bread
¼ pound butter
¼ pound butter, soft
2 tablespoons prepared horseradish
80 thin, small slices of sausage
2 onions, finely chopped

Cut off the crust and divide each slice of bread into 4 squares. Fry in hot butter in small portions until golden. Cool. Combine soft butter with horseradish. Spread over the toast. Place 2 slices of sausage on each canapé, sprinkle with onions.

YIELDS 40 CANAPÉS.

Spanish Canapés with Bananas

People living in southern Europe and in Latin America use more fruits in their meals and hors d'oeuvres because fruits are so abundant, flavorful, and inexpensive the year round, while meat spoils quickly in hot weather. Spaniards especially love fruits and vegetables, and thanks to that, there are not too many overweight senoritas.

40 thin slices long, narrow stale French bread
½ cup Mayonnaise*
40 slices tomatoes
Salt
40 slices bananas
2 tablespoons lemon juice

Spread bread with Mayonnaise. Cover with tomato slices. Sprinkle with salt. Garnish with bananas sprinkled with lemon juice.

YIELDS 40 CANAPÉS.

· ·

Danish Egg Spread

Danish open sandwiches are famous all over the world. They are the size of a slice of French bread, and similar bread is generally used. They are buttered and topped—or rather loaded—with everything you can imagine—all kinds of seafood, meats, salads, and spreads. Quite often egg spread is used under fish or ham. One of the restaurants specializing in these sandwiches, which I visited in Copenhagen, has a menu of 300 varieties listed on a six-foot-long strip of paper.

2 hard-boiled eggs, finely chopped
½ pound sweet butter, very soft
Salt

Mix the eggs with butter very well. Add salt to taste. The spread should not be very salty when salty meats or fish will be used to cover it.

YIELDS 1½ CUPS OF SPREAD.

. . . .

French Beef Spread

French girls are not only elegant and sexy, but thrifty housekeepers as well. This is how they use leftover beef. Remnants of boiled soup cuts, pieces of roast, stews, each separately or all together may be turned into a tasty spread for crackers or small toasted slices of bread.

1 medium dill pickle
3 green onions
1 small tomato
1 cup leftover beef, ground
2–3 tablespoons strong bouillon from a cube
Salt and pepper
1 tablespoon green parsley, chopped

Chop the pickle, onions, and tomato very fine. Combine with meat. Add enough bouillon for easy spreading. Season, mix well.

Spread generously over crackers; sprinkle with parsley.

YIELDS 1½ CUPS OF SPREAD.

. . . .

French Chicken Liver Pâté

It is nice to have all kinds of little snacks and canapés for a cocktail party, but we do not always have time to prepare them. A much easier way is to serve two or three kinds of good cheese, a tempting, famous French pâté in a dish, plenty of crackers—and let the guests help themselves.

1 pound chicken livers
1 cup water
¼ cup minced onions
½ pound butter, soft
2 teaspoons dry mustard
2 teaspoons salt
¼ teaspoon cloves
¼ teaspoon pepper
3 tablespoons sherry

Bring the livers to a boil, cover, and simmer for 20 minutes. Drain and grind twice, along with the minced onions. Blend all the ingredients in a mixer. Place in a dish in form of a ball.

SERVES 20.

Italian Pork Pâté

The French like their pâtés mostly from poultry liver, especially goose liver. Italians ventured into preparing it from pork, and the result is delightful. Ask your butcher to grind your meat.

Cook it a day or two before your party. Serve with crackers, and let your guests help themselves.

¾ pound pork liver, ground twice
½ pound pork, ground twice
½ pound lard, soft
Salt and pepper to taste
½ teaspoon nutmeg
½ teaspoon paprika
1 onion, very finely chopped
1 tablespoon butter
2 small eggs, lightly beaten
2 tablespoons flour

Combine liver and pork with lard. Season well with salt, pepper, nutmeg, and paprika. Sauté onions in hot butter. Combine with the meat and eggs mixed with flour, and beat all ingredients very well in a mixer. Cook for 1½ hours in a double boiler that had been generously spread with lard and sprinkled with bread crumbs.

Cool thoroughly. Remove from the pan.

YIELDS ONE 1½ POUNDS PÂTÉ.

Steak Tartare—A European Fad

This is a dish of raw beef, the latest favorite of cocktail party fans. Centuries ago it was on holiday menus of Tartar tribes in the Turkish empire. Later it was considered barbaric and forgotten for a long while. During the last decade, it returned to popularity in France, and soon caught the fancy of German, Scandinavian, and East European gourmets.

Steak Tartare is served on crackers garnished with capers or green parsley or on small plates as an after-theater snack along with a bowl of salad. It is considered sophisticated and good for the waistline, but it is wise always to have a choice of something else for guests who do not care for raw meat.

Choose a lean, unmarbled piece of sirloin. Ask the butcher to trim off all the fat and to grind the meat twice. Buy the meat on the same day you plan to serve it.

> 2 pounds twice-ground steak
> 2 teaspoons salt
> ½ teaspoon freshly ground pepper
> ¼ cup finely chopped onions
> 4 raw egg yolks with white threads carefully removed
> 2 tablespoons finely chopped green parsley

Mix the meat with salt, pepper, onions, and egg yolks. Refrigerate for 1 hour. Spread generously on crackers. Sprinkle with parsley.

TO SERVE AS HORS D'OEUVRES: Mix the meat with salt, pepper, and onions, add 2 teaspoons Worcestershire sauce. Refrigerate. Divide into 6 portions. Form 6 steaks on six small plates. Make a hollow and place a whole raw egg yolk in each one. Sprinkle with parsley.

SERVES 6.

Toasted Cheese Canapés from Hamburg

Hamburg, the large German Baltic harbor full of visiting merchants and seamen, is known for its nightlife. Housewives in Hamburg often have to entertain international guests, their husbands' business associates and customers, at a moment's notice. They treasure their quick recipes for elegant snacks.

It takes 15 minutes to prepare these canapés.

6 slices white bread
1 ¾ cups grated cheese
6 tablespoons white table wine
¼ cup coffee cream
¼ teaspoon paprika

Cut each bread slice in half. Toast in a 450° oven for 8 minutes.

Combine cheese with wine, cream, and paprika. Spread on toasts. Broil a few minutes until golden and bubbling.

SERVES 4.

. . . .

French Grilled Canapés with Wine

It seems that in France you can find wine everywhere and in everything, even in bread, or as in this case, soaked into small sophisticated finger snacks that taste so good with cocktails.

40 thin slices long, narrow French bread
1 cup white table wine
1 cup grated Parmesan cheese
2 eggs, slightly beaten
¼ cup butter, melted

Dip each slice of bread in wine. Combine the cheese with the egg. Spread over the canapés. Arrange on a buttered cookie sheet. Sprinkle with butter. Bake at 450° for 10 minutes or until golden brown. Serve warm.

YIELDS 40 CANAPÉS.

. . . .

Dill Strudel from Prague

Hungarian strudel dough was appropriated by cooks of many nations. Available in refrigerated sections of many supermarkets and gourmet stores, it is useful in preparing different kinds of snacks. In Czechoslovakia strudel is often served as an hors d'oeuvre with drinks or before dinner.

Strudel dough dries rapidly, so do not open the package before the filling and everything else is prepared. Work fast.

2 sheets strudel dough
¼ cup butter, melted

STUFFING
1 pound farmer cheese
3 egg yolks
¼ cup sour cream
¼ teaspoon salt
2 teaspoons dill weed
1 egg white

TO MAKE STUFFING: Press the cheese through a strainer. Combine with egg yolks, sour cream, salt, and dill. Beat the egg white until stiff. Fold into the cheese mixture.

TO MAKE STRUDEL: Spread a damp large cloth napkin on a table. Unfold strudel dough. Place one sheet on the napkin. Brush with half of the butter. Place the second sheet over it. Brush with the remaining butter.

Place the stuffing along the short edge in a 4-inch strip. Roll with the aid of the napkin. Bake on a buttered cookie sheet in 400° oven for 30 minutes.

Before serving reheat in a 350° oven for 10 minutes.

SERVES 10.

· ·

Meat and Mushroom Strudel from Prague

It is nice to have at least one warm snack at a cocktail party even if you didn't plan a buffet. You can prepare your stuffing a day ahead, bake your strudel in the afternoon, and reheat it in a moderate oven just before serving.

Two sheets of strudel dough or filo
½ cup butter, melted

STUFFING
1 large onion, finely chopped
½ pound mushrooms, finely chopped
3 tablespoons bacon drippings
2 cups leftover roast, ground
Salt and pepper

TO MAKE STUFFING: Fry the onions and mushrooms in hot fat until golden. Add the meat and fry for 5 more minutes, stirring. Season with salt and pepper.

TO MAKE STRUDEL: Spread a damp large cloth napkin on a table. Unfold strudel dough. Place 1 sheet on the napkin. Brush with half of the butter. Place the second sheet over it. Brush with the remaining butter. Place stuffing along the short edge in a 4-inch strip. Roll with the aid of the napkin. Place on a buttered cookie sheet in 400° oven and bake for 30 minutes.

SERVES 8.

· ·

. . . .

Sausage Strudel from Bratislava

On two sides of the Carpathian Mountains live two nations, the Slovaks and Poles, closely related by blood, language, and customs, but nurtured by history in two different
ways.

This dish shows how the Slovaks stretch one hand toward
Hungary, and the other to its Polish neighbor on the north.
Both influences combine nicely in Hungarian strudel dough
with Polish sausage.

1 pound Polish sausage, skinned, ground
¾ cup grated cheese
⅓ cup bread crumbs
1 egg
2 sheets strudel dough
¼ cup bacon drippings, melted

TO MAKE STUFFING: Combine sausage with cheese,
bread crumbs, and egg.

TO MAKE STRUDEL: Roll and bake as in Meat and Mushroom Strudel from Prague.*

SERVES 10.

. . . .

Swedish Meatballs

There are two ways of serving Swedish meatballs. They
can make a nice snack for a cocktail party, formed into very
small balls, about ¾ inch in diameter, and served on a tray with
toothpicks on the side. When you prefer to use the meatballs

as a buffet dish, they may be slightly larger, 1 inch in diameter, and should be served in gravy.

1 large onion, finely chopped
3 tablespoons butter
1 cup mashed potatoes
⅓ cup bread crumbs
⅓ cup coffee cream
1 egg
1 teaspoon salt
1 tablespoon dill weed
1 pound ground round (beef)
2 tablespoons oil

GRAVY
Pan drippings
1 tablespoon instantized flour
¾ cup light cream
Salt and pepper

TO MAKE MEATBALLS: Sauté the onions in 1 tablespoon butter on medium heat until golden. Combine with potatoes, bread crumbs, cream, egg, salt, dill, and meat. Knead for 5 minutes until smooth. Form into small balls. Arrange in one layer on a tray. Cover and refrigerate for 1 hour.

Heat the remaining butter with the oil. Fry meat balls 10 at a time on medium heat on all sides until brown, shaking the skillet to prevent sticking.

Keep warm in a 200° oven until needed, but not longer than 15 minutes. They may be reheated the same way or in the gravy.

TO MAKE GRAVY: Degrease the pan juices by skimming off the fat; stir in the flour and the cream. Bring to a boil and simmer for 3 minutes. Pour over the meatballs and serve.

SERVES 6.

Buffet Entertaining

PREFACE

A buffet is simply a Swedish smorgasbord—a way of serving food that proved so convenient that it has been adopted everywhere in Europe and is taking an increasingly important place in American social life.

Today young hostesses tend to favor informal entertaining. They seldom have a dining room large enough to seat all their guests, and they do not have or care to use all the dishes needed for the traditional setting of a dinner table. A buffet is a much simpler way of dealing with larger groups. I begin planning a buffet when the guest list reaches ten.

In Stockholm, Paris, Rome, or Berlin, most people live

in apartments. They seldom have houses as spacious as ours in America. And in this country too there is a growing trend to carefree apartment living, especially among the younger generation. However, this should not prevent pleasant evenings at home with friends.

An outstanding asset of a buffet is its flexibility. You can set it indoors or outdoors, in the living room, dining room, game room, or on a porch. You can have it at midday or in the evening. I have seen lovely afternoon Christmas buffets at which only pastries, punch, and wine were served. But even aside from convenience in terms of space and equipment, a buffet dinner has many other advantages. There is much more mingling among the guests. No one is stuck with a dinner partner he finds boring. Everyone can choose his companions for conversation, and also change them whenever he wishes. A buffet is certainly more fun and more of an adventure than a sit-down dinner party.

Party time is time for laughter, music, dance, and of course good food. It is pure joy to play and be merry when one is young and light-hearted, when one is with friends and can share one's happiness with them.

In Europe people try to make every party special with careful preparations and by dressing up. It is customary to bring your hostess a few selected flowers, a box of chocolates, or a bottle of wine. More worldly and wealthy Europeans send flowers on the morning of the party or a day after with a note of thanks.

The southern Europeans—the Spanish or the Italians—are hopelessly late for every social occasion, but in Norway it is taken as a great offense if the guest does not appear in the door at the very moment when the clock strikes the exact time. Poles, Czechs, and Hungarians are more relaxed about punctuality. They might be half an hour late for cocktails or an afternoon coffee, but they try hard to be on time for a dinner invitation.

Invite friends with common interests to your buffet party and include a few persons outside your usual crowd. One of the refreshing aspects of a party is the opportunity it offers to meet new people. You can have up to twenty guests for a buffet in a small house or apartment without worrying whether you have enough comfortable chairs for all of them. You do not worry whether your dinner plates or silver are uniform. Your guests will sit wherever they can and eat with their plates on their knees, or they may eat standing.

The practical hostess likes to plan a menu that can be prepared well in advance and served cold or reheated quickly. She can enjoy her guests and her party more that way. A perfect menu for a buffet requires very little cooking on the day of the party. Salads are excellent for this purpose because, with the exception of the fresh greens, most needed ingredients can be prepared a day ahead. Combining everything takes just a few moments. And buffet food need not be expensive. Economical, home-made casseroles are usually more appreciated than cold cuts from the store. Buffet entertaining is really the best way you can feed a group at surprisingly little cost.

Since your guests will not be sitting at the table, it is important to choose buffet dishes that do not require a knife or which involve very little cutting. A real Swedish smorgasbord includes 50 or more traditional delicacies, carefully decorated, and arranged on a long table. But a practical and modern classic buffet dinner consists of several salads, perhaps some open sandwiches, a pâté, and a good selection of meat dishes, casseroles, baked fish, and stews. Soup should never be offered at other than seated affairs. On a warm summer day the whole buffet may be cold, but in the winter time it is good to provide at least one or two hot dishes, such as a roast (it doesn't need much more attention than setting the stove's thermostat at the right temperature), and a dish that has been reheated.

All cakes from the dinner party chapter are suitable for a buffet. Tortes are simple to prepare in a mixer and always elegant. It is customary to cut small pieces, so one nice torte will feed a crowd. I bake my tortes a week before a party and freeze them.

Flowers on the table add enormously to the beauty of your buffet. In Europe people often eat hamburger instead of a steak, but they buy flowers for their table. When flowers are too expensive, you can use a decorative fruit bowl, an arrangement of gourds and corn in the fall, some dried weeds, around Christmas a few branches of pine or a centerpiece of cones—anything your imagination dictates. Candles also help in creating a festive mood. All girls look pretty and romantic in candle light.

There is no better way to show off your cooking and decorating talents than with a buffet. The food must be arranged on the table in an attractive, colorful display and in the proper order, beginning with seafood and salads. It is not advisable to use paper dinner plates as they are not sturdy enough, but you can certainly use paper napkins, and paper plates for the dessert. You can also buy inexpensive, attractive clear plastic glasses for all the drinks, even champagne. If you want your buffet to look nice, however, it is worth the trouble of laundering a nonplastic tablecloth later. You can find a lovely tablecloth that never requires ironing.

Clear the food from the table and remove the dinner plates before serving coffee and tea. Don't worry if your home looks a little messy during a party. It is often more fun that way. When the hostess strives for perfection, the atmosphere easily becomes stiff and her party is likely to be a failure.

Suggestions for Elegant Buffet Menus

DRESSINGS, RELISHES, HORS D'OEUVRES

1. French Mayonnaise
2. French Mayonnaise with Gelatin
3. Sauce Tartare from Brussels
4. French Sauce Vinaigrette
5. Cantaloupe from Avignon
6. Polish Red Beet Relish
7. Chantilly Asparagus
8. Stuffed Eggs from Strasbourg
9. Stuffed Cucumbers from Sweden
10. Stuffed Tomatoes from Valencia
11. Stuffed Tomatoes from Lake Balaton
12. Copenhagen Eggs Stuffed with Cheese
13. Fillets of Sole in Aspic from Marseilles
14. Continental Chicken in Aspic
15. Veal in Aspic from Finland
16. French Turkey and Ham in Aspic

SALADS

17. Continental Tomato Salad
18. Green Pepper Salad from Budapest
19. Polish Red Cabbage Salad
20. French Cauliflower Salad
21. Vegetable Salad from Luxembourg
22. Neapolitan Egg Salad
23. Russian Vegetable Salad
24. Potato and Lettuce Salad from Budapest
25. Potato Salad from the Petit Trianon
26. Herring Salad from Helsinki
27. Jewish Herring Potato Salad

28. *Mixed Salad from Lake Balaton*
29. *Mushroom Salad from Germany*
30. *Mushroom Salad from Florence*
31. *Danish Salmon Salad*
32. *Turkey Salad from Versailles*
33. *Russian Vegetable Ham Salad*
34. *Beef Salad from Moscow*

HOT BUFFET DISHES

35. *Hungarian Cabbage Strudel*
36. *French Asparagus in Cheese Sauce*
37. *Baked Sole the Norwegian Way*
38. *Turkey with Almond Stuffing from Prague*
39. *Turkey with Raisins from Vienna*
40. *Baked Ham Continental Style*
41. *Spaghetti with Sausage the Bulgarian Way*
42. *English Roast Beef*
43. *Hungarian Goulash*
44. *Marjoram Goulash from the Great Plain*
45. *Rice Casserole Valencia Style*
46. *Transylvanian Lamb Casserole*
47. *Italian Pork Casserole*
48. *Pork and Cabbage from Helsinki*
49. *Hungarian Sauerkraut Goulash*
50. *Lithuanian Hunters' Stew*

. . . .

French Mayonnaise

It is truly impossible to imagine a European party without French Mayonnaise. This dressing adorns the famous canapés of Denmark, as well as any smorgasbord table in Sweden, forms an integral part of classical Russian salads, and decorates elegant hors d'oeuvres in Warsaw, Budapest, and Vienna. It is so popular simply because of its incomparable taste. And the beauty of it all is not just in the pale ivory color of this favorite sauce of the French, but in the fact that it is so easy to prepare. It is difficult to believe, but in five minutes you can have a whole jar of true French Mayonnaise, so much better than any of the commercial ones you can find in supermarkets.

1 egg yolk
1 tablespoon lemon juice
1 cup salad oil
3 tablespoons water
¼ teaspoon salt
¼ teaspoon sugar

Beat the egg yolk in the mixer at low speed for 1 minute. Still beating, add the lemon juice in 4 portions, then the oil in a very thin stream, and the water by teaspoons. Season. Keep refrigerated.

YIELDS 1 CUP.

. . . .

French Mayonnaise with Gelatin

Molded salads are always elegant. A molded salad in Mayonnaise* adds a touch of originality to your menu.

This sauce is also often used by practical French hostesses who like to cover their hors d'oeuvres with Mayonnaise well in advance of serving, but are afraid that it may slip to the bottom of the dish; gelatin successfully prevents this from happening.

1 tablespoon plain gelatin
2 tablespoons cold water
*1 cup Mayonnaise**

Soak gelatin in water for 5 minutes. Heat in a double boiler, stirring until completely dissolved. Mix with Mayonnaise. Immediately combine with your salad or pour over hors d'oeuvres.

YIELDS 1 CUP.

Sauce Tartare from Brussels

Sauce Tartare is famous all over the world, but in each country it is somehow varied and used in a different way. It is not necessarily served with seafood. In Brussels, for example, it is used mostly over cooked vegetables and cold roasts.

2 egg yolks
2 hard-boiled egg yolks
1 tablespoon lemon juice
1 cup salad oil
2 tablespoons water, boiling
¼ teaspoon salt
¼ teaspoon sugar
1 tablespoon prepared mustard
1 tablespoon capers, very finely chopped (may be omitted)
1 tablespoon dill pickles, very finely chopped
½ teaspoon green parsley, finely chopped
½ teaspoon basil

Beat the egg yolks in the mixer at low speed for 1 minute. Add the lemon juice in 4 parts, beating all the time. Add the oil in a very thin stream, and water by teaspoons, beating constantly. Combine with the remaining ingredients.

YIELDS 1½ CUPS.

French Sauce Vinaigrette

Your waiter in a restaurant in France would be extremely surprised if you asked for a French dressing on your salad. Many different dressings are used in France and all of them are equally French. Sauce Vinaigrette is one of most popular for simple green salads, tomatoes, and cucumbers. It takes five minutes to prepare and it certainly tastes better and costs only a fraction of what you have to pay for a ready-made dressing. Use the herbs you like most (such as parsley, dill weed, basil).

1 tablespoon vinegar
4 tablespoons oil
Salt, pepper, aromatic herbs

Combine all the ingredients. Beat with a fork for 5 minutes to blend very well.

SERVES 6.

Cantaloupe from Avignon

We are used to serving cantaloupe for breakfast or before meals. In Avignon they eat it in a different way.

In 1309, when the Pope went to live in the south of France, he took with him a Roman variety of melon. The French named it after Cantalupo, then the papal residence on the outskirts of Rome. People in Avignon learned to cultivate Roman melons, but they did not like them too much at the beginning, and so they tried to improve the taste by marinating them with port, and served them as a relish with different meats.

A good cantaloupe is heavy. The skin gives a little under pressure, and the aroma is strong.

1 cantaloupe
1 cup port

Cut the top off the cantaloupe and reserve. With a long spoon remove all the seeds. Pour port inside the cantaloupe and replace the top. Let stand for 2 hours.

Drain the juice and save. Slice and peel the cantaloupe. Arrange on a serving platter. Sprinkle with the juice. Chill.

SERVES 10.

. . . .

Polish Red Beet Relish

This is a traditional Polish relish served with ham, sausage, and other cold meats at the Easter Sunday midday feast. Its lovely color adds life to any buffet table. Serve in a soup bowl with a teaspoon, which should indicate that it is hot and should be used sparingly.

You may keep this tasty relish refrigerated in a covered jar for several weeks.

2 1-pound jars canned red beets
6 ounces prepared horseradish
2 teaspoons sugar

Drain and chop the red beets. Combine with horseradish and sugar. Return to the jars; cover. Refrigerate overnight.

SERVES 15.

. . . .

Chantilly Asparagus

The medieval palace of Chantilly, situated not very far from Paris, was the home of aristocratic beauties and the kings' seductive mistresses. Reconstructed in the 16th century, embellished by Louis XIV and XV, the splendid palatial rooms witnessed many enchanting receptions at which exquisite food was served. Chantilly also became famous for its hand-made lace—delicate as morning fog—as well as for whipped cream and sauces light and foamy as the lace. Asparagus with whipped cream sauce is a favorite vegetable in Chantilly.

3 pounds asparagus tips, cooked
2 cups Mayonnaise*
4 tablespoons whipped cream

Cook asparagus in the afternoon. Make Mayonnaise thick, adding very little water. Arrange asparagus on a serving dish, cover, and refrigerate.

BEFORE SERVING: Mix Mayonnaise with whipped cream, pour over the asparagus.

Broccoli cut into long thin spears may be served the same way.

SERVES 12.

. . . .

Stuffed Eggs from Strasbourg

Strasbourg in the French province of Alsace is not a very large city, but is quite industrial. It has a lovely old cathedral and well-preserved romantic medieval buildings. It must be also blessed with imaginative people because many famous dishes bear its name.

3 chicken livers, cut up
2 tablespoons salad oil
3 medium mushrooms, sliced
6 hard-boiled eggs, cut into halves
1 cup Mayonnaise with Gelatin*
Parsley sprigs

Sauté chicken livers in hot oil. Sauté mushrooms in a separate pan. Grind the livers with the egg yolks. Stuff the egg whites. Place mushroom slices on the top of the stuffing. Place the eggs on a serving platter. Cover with Mayonnaise. Garnish with parsley.

SERVES 12.

. . . .

Stuffed Cucumbers from Sweden

Cucumbers are very much liked in Scandinavia and are used in many ways, much more often than is customary here. They are served in salads and casseroles; they are boiled, they are baked, and they are stuffed. When in season, they appear on the Swedish table several times a week.

Stuffed cucumbers make a very decorative, original dish on a buffet table.

2 large cucumbers peeled
½ cup chopped radishes
½ cup grated carrots
2 tablespoons green parsley, finely chopped
1 tablespoon dill weed
One 4-ounce can crab meat, finely chopped
1 cup Mayonnaise*
1 cup sour cream
Salt and pepper
1 pint cherry tomatoes, cut into halves

Scratch the sides of cucumbers lengthwise with a fork to make grooves. Cut each cucumber into 1-inch-thick slices. Hollow each slice half way through.

Combine radishes with carrots, parsley, dill, and crab meat. Combine mayonnaise with sour cream. Add to the salad and season.

Stuff cucumber slices with the mixture. Place a tomato half at the center of each stuffed cucumber slice.

SERVES 12.

. . . .

Stuffed Tomatoes from Valencia

When you hear the word "Spain" you think of handsome bullfighters and lovely girls in wide colorful skirts with lace headdresses. Valencia is famous for its lace but no less for tasty dishes made of fish and vegetables.

12 small tomatoes
2 cups cooked fish, ground
2 tablespoons green parsley, chopped
4 hard-boiled eggs, quartered
¾ cup Mayonnaise*
12 anchovy fillets, finely chopped
Lettuce leaves, finely shredded

Cut off the tops of tomatoes and remove central pulp. Combine fish with parsley. Stuff tomatoes. Place ¼ egg on the top of each. Combine Mayonnaise with anchovies. Spoon over the egg.

Line a serving platter with lettuce and arrange tomatoes on it.

SERVES 12.

. . . .

Stuffed Tomatoes from Lake Balaton

The shores of the Lake Balaton are a real Hungarian Riviera where the rich used to have their summer villas. Many lovely swimming and boating resorts accommodate thousands of vacationers from all over the country and numerous foreign visitors. On summer evenings the sweet music of violins resounds over the water while people dine in romantic Hungarian taverns.

This recipe is a good way of using leftover fish. Prepare your tomatoes the afternoon of the party. Cover tightly with plastic and refrigerate.

12 small tomatoes
1 pound cooked cold fish
1 pint Tartar sauce
Salt and pepper
Parsley sprigs

Cut off the tops of tomatoes and remove central pulp. Place them upside down to drain. Pass the tomato pulp and juice, fish, and half of the Tartar sauce through a strainer. Mix well.

Sprinkle the interior of the tomatoes with salt and pepper. Stuff; cover with tops. Place on a serving platter. Spoon over the remaining Tartar sauce. Decorate with parsley.

SERVES 12.

Copenhagen Eggs Stuffed with Cheese

The Danes are great farmers, known for the excellence of their dairy products. Cream and cheese are served there more often than elsewhere. The last time I was in Copenhagen, I stayed at a small hotel patronized mostly by Scandinavian traveling teachers and missionaries. The cuisine was simple but truly Danish and very tasty. These eggs were served as part of a Sunday night smorgasbord.

5 hard-boiled eggs, cut into halves
2 tablespoons butter, soft
3 tablespoons grated Parmesan cheese
3 radishes, sliced

Press the egg yolks through a sieve. Combine with butter and cheese. Stuff the eggs. Place slices of radishes on the top of each egg.

SERVES 10.

Fillets of Sole in Aspic from Marseilles

The warm climate of France's most famous and romantic harbor city calls for many cold dishes. Fish in aspic is a pleasant surprise at every buffet. This is how they make it in France.

3 pounds fillets of sole
2 cups white dry wine
1 envelope plain gelatin
2 tablespoons cold water
¼cup Mayonnaise*

Simmer fish in wine for 20 minutes. Soak gelatin in cold water for 10 minutes.

Carefully remove the fish to a serving platter. Add gelatin to the hot fish broth; stir until dissolved. Pour over the fish. Refrigerate. Decorate with Mayonnaise before serving.

SERVES 12.

Continental Chicken in Aspic

Chicken in aspic is easy to prepare, tasty, and inexpensive; a perfect dish for every buffet. Your aspic will always be nice and clear if you cook your chicken very gently at the lowest possible heat so that the broth is barely boiling.

Two 3-pound broilers
1 quart water
2 celery stalks
1 carrot
1 parsnip
1 large onion, quartered
2 teaspoons salt
1 bay leaf
4 peppercorns
1 teaspoon tarragon
½ cup shredded lettuce
6 slices lemon

Place chickens in a kettle with water. Add vegetables and spices; bring to a boil. Reduce the heat and cook very gently for 2 hours. Cool slightly. Skin and bone the chickens, cut into 1-inch pieces. Strain the broth. Refrigerate, removing the crust of fat after broth has chilled. Heat until liquid.

Spread a pretty mold with salad oil. Pour in 1 cup of cold broth. Refrigerate until set. Pour in the rest of the broth combined with chicken. Refrigerate until set.

Before serving, unmold by loosening edges with paring knife and shaking onto a chilled serving plate. Garnish with lettuce and lemon.

SERVES 12.

Veal in Aspic from Finland

Finnish women do not need a Women's Lib movement. They have been liberated for a long time. They were the first

(1907) in Europe to get the right to vote. Since then the doors
of all the nation's universities have been open to them. There
are many successful professional women in Finland, and they
are believed to be as good physicians, laywers, engineers, and
executives as the men are. This does not prevent them from
appreciating the art of cooking, however.

Try this typical Finnish smorgasbord dish, which can be
prepared a day ahead of your party.

2 pounds shoulder veal, cut into large pieces
2 cups boiling water
8 peppercorns
2 carrots, cut into halves
1 bay leaf
Salt to taste
1 ½ tablespoons vinegar
2 canned beets, sliced
½ cup sour cream

Add the meat to boiling water; add peppercorns, carrots,
and bay leaf. Bring to a boil uncovered. Skim off the scum and
season with salt. Cover and simmer for 1 hour. Cool. Cut the
meat into small cubes. Drain the stock and add vinegar.

Spread a mold with salad oil. Pour in ½ cup cold stock.
Refrigerate until set. Arrange carrot slices on the aspic. Place
the meat on it. Pour the remaining stock. Refrigerate overnight.

Unmold by loosening edges with a paring knife and shaking onto a chilled serving plate. Garnish with beet slices.
Decorate with sour cream.

SERVES 10.

French Turkey and Ham in Aspic

Thrifty French housewives know a million tricks for serving leftovers in most delectable ways. This dish is extremely convenient to prepare for a party in the few days after Christmas when you don't know what to do with all the turkey on hand.

2 cups leftover turkey, cut up into small pieces
1 cup ham, coarsely chopped
2 hard-boiled eggs, coarsely chopped
1 medium dill pickle, finely chopped
1 envelope plain gelatin
2 tablespoons cold water
1½ cups strong chicken bouillon
½ cup white dry wine
1 tomato, sliced

Combine turkey with ham, eggs, and pickles. Soak gelatin in water for 5 minutes and add to hot broth; stir until dissolved. Add wine. Add liquid to the turkey mixture. Cool. Pour into an attractive mold that has been greased with salad oil. Refrigerate until set.

Unmold by loosening edges with paring knife and shaking onto chilled serving plate. Garnish with tomatoes.

SERVES 10.

Continental Tomato Salad

This is the way tomatoes are usually served on the Continent. Fastidious European cooks recommend dipping each tomato in boiling water for a minute and peeling it before slicing. We don't often do it that way any more.

Prepare your tomatoes in advance, arrange on a serving platter, and refrigerate under cover. Don't sprinkle them with the dressing until just before serving. Let them sit at room temperature for half an hour before your buffet is ready. Continental gourmets believe that vegetables are not at their best when very cold.

6 medium tomatoes, thinly sliced
½ cup Sauce Vinaigrette*
1 tablespoon green parsley, finely chopped
2 medium onions, thinly sliced

Sprinkle tomatoes with the sauce. Sprinkle with parsley; garnish with onions.

SERVES 12.

. . . .

Green Pepper Salad from Budapest

What is one man's delight is another man's poison. To Hungarians, green peppers are always delicious. We certainly may benefit by imitating them since green peppers are not only tasty, but very rich in vitamins.

This is a superb, simple salad for hot days in late summer when peppers are in season and at their best, and we don't crave hot dishes.

12 green peppers
Salt
¼ cup salad oil
2 tablespoons vinegar
1 teaspoon sugar
⅛ teaspoon pepper

Wash the peppers, remove seeds and inside ribs, slice. Throw them into salted boiling water. Boil for 1 minute. Drain in a colander. Make a dressing of the remaining ingredients. Mix with the peppers. Let stand for 1 hour.

SERVES 12.

. . . .

Polish Red Cabbage Salad

Most precious to a successful hostess are recipes for vegetables and salads that can be prepared well in advance. They are always in demand as there are not too many of them. Red cabbage salad has a refreshing flavor and will also make your buffet colorful. Prepare it on the night before your party. It complements all roasts, fried meats, and fish. Serve well chilled.

One 1 ½-pound head red cabbage, coarsely shredded
1 teaspoon salt
¼ cup lemon juice
3 tablespoons sugar
2 large apples, peeled, coarsely shredded

Cover the cabbage with boiling water. Bring to a boil. Cook over high heat for 7 minutes. Drain. Combine with the remaining ingredients. Refrigerate for several hours.

SERVES 8.

. . . .

French Cauliflower Salad

The famous refinements of French cooking are largely due to Catherine de Medici, the Italian wife of Henry II. She brought with her from Florence the custom of serving vegetables. Before the arrival of her court, even on the tables of the French nobility one could find only roasted and boiled meats, some fish, and a little honey for dessert.

> 1 medium cauliflower, divided into small flowerets
> 4 medium potatoes, boiled in jackets, cooled, peeled, sliced
> 1 cup Mayonnaise*
> 2 cups red cabbage, thinly sliced
> ½ cup Sauce Vinaigrette*
> 2 hard-cooked eggs, sliced
> Lettuce leaves

Place cauliflower in boiling salted water, cook for 15 minutes, drain, cool. Combine with potatoes and Mayonnaise. Cover red cabbage with boiling, salted water. Cook for 10 minutes, drain, cool. Combine with Sauce Vinaigrette.

Pile cauliflower salad in the middle of a serving platter. Cover with egg slices. Place a few lettuce leaves on the top. Arrange red cabbage around the cauliflower salad.

SERVES 10.

Vegetable Salad from Luxembourg

The smallest of the nations in the world (999 square miles) gained fame for its light and crisp Moselle wines, which are so good with salads. Old castles and fortresses are perched among steep hills and narrow valleys covered with oak and pine forests, making Luxembourg look like a fairyland country.

People in Luxembourg do not mix their salads, but arrange vegetables and meats in attractive circles, and then sprinkle them with Sauce Vinaigrette.

1 medium cauliflower, divided into small flowerets,
   cooked
3 tomatoes, cut into halves, thinly sliced
2 cups green string beans, cooked, coarsely chopped
*½ cup Sauce Vinaigrette**
2 hard-boiled egg whites, finely chopped

Arrange cauliflower in a small heap in the middle of a serving platter, surround with tomato slices, and then make a circle of beans. Sprinkle everything with the dressing. Top the cauliflower with egg whites.

SERVES 8.

. . . .

Neapolitan Egg Salad

Italians love *pasta,* and a true Italian salad quite often is mixed with macaroni. Serve this one with ham or other cold cuts. It will make a simple-to-prepare, inexpensive treat especially appropriate for summer evenings.

2 cups elbow macaroni, cooked
3 packages frozen mixed vegetables, cooked, drained
*1 cup Mayonnaise**
3 hard-boiled eggs, quartered
2 medium dill pickles, sliced
4 tomatoes, sliced
Salt and pepper

Combine macaroni with vegetables and Mayonnaise. Arrange on a serving platter in a small heap. Place the eggs and pickles over it in a decorative way. Arrange a circle of tomatoes around the salad. Sprinkle tomatoes with salt and pepper.

SERVES 12.

. . . .

Russian Vegetable Salad

The fame of this vegetable salad is by no means limited to Russia. It is well liked and served often in Paris and has been adopted by the best restaurants in all of Central Europe.

You will find here a characteristic seasoning—dill weed —that is highly popular in Continental cookery, especially in Scandinavia, Russia, Poland, and Hungary. Dried dill weed is easily obtained in the United States in larger supermarkets or the gourmet sections of department stores.

Serve this salad with a roast or cold cuts.

8 medium potatoes
2 1-pound cans mixed vegetables, drained
1 1-pound can peas, drained
4 medium dill pickles, diced
2 tablespoons dill weed
Salt and pepper
¾ cup Mayonnaise*
¾ cup sour cream
2 tablespoons prepared mustard
4 hard-boiled eggs, quartered
Lettuce leaves

Cook the potatoes in their skins. Cool, peel, and dice. Mix all the vegetables, add dill, salt, and pepper. Mix Mayonnaise with sour cream and mustard, add to the salad. Chill for several hours. Put into a dish; garnish with lettuce. Arrange eggs on the top and serve.

SERVES 10.

Potato and Lettuce Salad from Budapest

Hungarian salads are made chiefly of potatoes and cooked vegetables as greens are available mostly only in season. But when summer comes, cucumbers and lettuce are often served with sour cream or added to popular salad recipes.

This salad is excellent with baked ham or sausage.

4 potatoes, boiled, sliced
1 can white beans, drained
1 large dill pickle, cubed
4 tomatoes, cubed
2 green peppers, seeds removed, cut into strips
1 medium head lettuce
2 hard-boiled eggs, sliced
1 pint Mayonnaise*
1 tablespoon green parsley, chopped

Combine potatoes with beans, dill pickle, tomatoes, and peppers. Remove the large leaves from the lettuce and line a serving dish. Cut the rest of the lettuce into thin strips. Combine the lettuce, eggs, and Mayonnaise with the salad. Arrange on lettuce leaves. Sprinkle with parsley.

SERVES 12.

Potato Salad from the Petit Trianon

The last queen of France, Marie Antoinette, who died so tragically under the guillotine, loved to play the make-believe life of a shepherdess. In her charming little Petit Trianon palace in Versailles, she often ordered for her guests simple provincial dishes that tasted delectable after long promenades and games in nearby meadows.

Serve with roasts or hot sausages.

8 medium potatoes, boiled in their jackets
1 cup Sauce Vinaigrette*
2 tablespoons green parsley, finely chopped
3 tablespoons green onions, finely chopped
1 cup white table wine, hot

Boil your potatoes until just tender. Peel and slice while warm. Toss gently with the sauce. Let stand 5 minutes. Sprinkle with parsley and onions. Refrigerate under cover.

Remove and allow to stand at room temperature for 1 hour before serving. Toss with hot wine and serve immediately.

SERVES 12.

Herring Salad from Helsinki

There is no formal Christmas dinner in Finland. People sleep late and like to relax in a family circle or to pay short visits to close friends. After the copious feast in the middle of the night on Christmas Eve, no one is really hungry, so leftovers and small dishes prepared in advance, like this herring salad, are usually served.

Two 8-ounce jars of herring in wine
2 dill pickles
2 cups canned diced beets, drained
1 cup potatoes, boiled
2 apples
Salt
2 hard-boiled eggs, chopped

DRESSING
1 pint sour cream
2 tablespoons prepared yellow mustard
½ teaspoon sugar
1 teaspoon vinegar
1 tablespoon beet juice
Salt to taste

Drain and coarsely chop herring fillets. Combine with diced vegetables, and apples. Season with salt. Mix the ingredients of the dressing. Combine with the salad. Arrange in a serving bowl. Sprinkle with egg.

SERVES 12.

Jewish Herring Potato Salad

Herring and carp are favorite fish in the Jewish cuisine. It is tedious to soak, skin, and fillet salted herring. But, luckily, prepared herring in wine sauce may be used successfully for many recipes. It is just right for a herring potato salad.

Two 8-ounce jars herring in wine sauce
3 dill pickles, finely diced
2 apples, peeled, cored, diced
6 medium potatoes, cooked, cooled, peeled, diced
*1 cup Mayonnaise**
3 tomatoes, cut into wedges

Drain herring and onions from the jars, saving the sauce. Cut herring fillets into ½-inch slices. Chop the onions finely. Combine with pickles, apples, and potatoes. Mix the Mayonnaise with a few spoons of the wine sauce. Combine with the salad. Transfer onto a serving dish. Garnish with tomato wedges.

SERVES 10.

. . . .

Mixed Salad from Lake Balaton

This is an unusual recipe combining mushrooms and green peppers browned in oil with fresh vegetables. It is especially suitable with Hungarian goulash or Hungarian chicken with paprika. Prepare everything in advance, but mix with lettuce, dressing, and seasonings just before serving.

12 ounces mushrooms, quartered
2 green peppers, sliced
½ cup salad oil
¼ cup wine vinegar
1 pound asparagus
2 tomatoes, cut into wedges
½ cucumber, peeled, cut into halves lengthwise and
*   sliced*
1 head lettuce, shredded coarsely
Salt and pepper to taste

Fry the mushrooms in hot oil until golden. Add green peppers, fry for a few minutes more. Remove mushrooms and peppers from the skillet and refrigerate. Add vinegar to the oil in the skillet, mix well. Cook asparagus in salted water for 10 minutes, drain. Cut off the hard sections of the stalks. Cut the remaining parts into 1½-inch pieces. Cool.

Combine the mushrooms and green peppers with asparagus, tomatoes, cucumber, lettuce, and the vinegar dressing. Season with salt and pepper.

SERVES 8.

Mushroom Salad from Germany

The famous creator of *Faust* greatly valued the pleasures of the table. Almost every one of his famous friends mentioned in his letters or memoirs that Goethe avoided culinary eccentricities, preferring simple bourgeois dishes. Mushrooms were served quite often in his home.

2 pounds mushrooms
Salt
1 teaspoon lemon juice
1 cup green onions, finely chopped.
*1 cup Mayonnaise**
1 cup lettuce leaves, shredded
2 hard-boiled eggs, sliced

Trim the ends, rinse the mushrooms thoroughly. Place in a saucepan, cover with salted boiling water. Cook for 5 minutes. Drain. Slice mushrooms and sprinkle with lemon juice. Let stand for 5 minutes. Combine with half of the onions and with the Mayonnaise. Refrigerate for 1 hour.

Arrange in a heap on a serving platter, sprinkle with the

remaining onions. Garnish with a circle of lettuce and place egg slices over it.

SERVES 10.

- -

Mushroom Salad from Florence

Italians were the first to venture into tasting raw mushrooms. The idea spread quickly to France, and now raw mushrooms are popular everywhere among the gourmets and amateurs of Continental delicacies.

Choose mushrooms fresh and white, rinse well in a colander without soaking. Cut off the ends of the stems.

4 ounces mushrooms, thinly sliced
Salt
Red pepper
1 tablespoon lemon juice
4 tablespoons salad oil
2 tomatoes, cut into halves, thinly sliced
1 large onion, thinly sliced
*½ cup Sauce Vinaigrette**
1 small dill pickle, sliced

Sprinkle the mushrooms with salt and pepper. Combine with lemon juice and oil. Arrange into a small heap in the middle of a serving platter. Surround with tomato slices. Arrange onion slices over tomatoes. Sprinkle tomatoes and onions with Sauce Vinaigrette. Place a pickle slice over each onion slice.

SERVES 6.

. . . .

Danish Salmon Salad

The Danes are known all over the world for their friend-liness and good humor. They are cheerful and light-hearted, but at the same time very industrious. That combinaton has made them not only prosperous but well liked and respected everywhere. Surrounded on three sides by the sea, Danes have always been good seamen and fishermen. Danish wives have learned to prepare many tempting fish dishes.

2 cups potatoes, boiled, cubed
One 1-pound can red salmon, drained, crumbed
1 cup Mayonnaise*
2 medium tomatoes, sliced
2 hard-boiled eggs
1 tablespoon green onions, finely chopped
1 tablespoon green parsley, finely chopped

Arrange potatoes on the bottom of a serving platter. Cover with salmon. Sprinkle with Mayonnaise. Cover with circles of tomato slices alternating them with the egg. Sprinkle with onions and parsley. Refrigerate for 1 hour.

SERVES 8.

. . . .

Turkey Salad from Versailles

The chefs of the splendid residence of the French kings contributed greatly to the elegance of the Continental way of entertaining. This salad, artistically decorated and served on silver platters, adorned many regal receptions in Versailles. It is simple and tasty, a perfect way of using leftover turkey.

2 cups leftover turkey, chopped
¼ head lettuce, finely sliced
3 medium potatoes, boiled in jackets, peeled, cooled, chopped
1 ¼ cups Mayonnaise*
2 tablespoons tomato purée
½ cup fresh horseradish, chopped in a blender

Combine turkey with lettuce, potatoes and 1 cup Mayonnaise. Arrange in the form of a dome on a serving platter. Spread with the remaining Mayonnaise mixed with tomato purée. Sprinkle with horseradish.

SERVES 6.

. . . .

Russian Vegetable Ham Salad

It is economical to save and freeze leftover lean pieces from a baked ham for later use in a salad. Do not serve ham salad at the same buffet at which you serve ham. The greatest attraction of a fine buffet is the variety of tastes it offers.

6 medium potatoes, cooked in skins, peeled, diced
2 packages frozen mixed vegetables, cooked, drained
2 medium dill pickles, diced
2 cups ham, coarsely chopped
2 tablespoons green parsley, chopped
Salt and pepper
1 ½ cups Mayonnaise*
2 tablespoons prepared mustard
2 hard-boiled eggs, sliced
3 canned beets, sliced
Lettuce leaves

Combine potatoes with mixed vegetables, pickles, ham, and parsley. Season with salt and pepper. Add Mayonnaise mixed with mustard. Arrange in form of a dome on a serving platter. Decorate with eggs and beets; garnish with lettuce leaves.

SERVES 12.

Beef Salad from Moscow

This salad is a perfect example of the strong influence French cuisine has had on European cookery. More French in character than Russian, this salad was as popular in Moscow before the Revolution as it is now. A real favorite, it is served with vodka in the best Russian hotels.

1 small cauliflower, divided into flowerets, cooked
4 stalks celery, cut into 2-inch pieces
1 cup green olives, pitted
1 cup black olives, pitted
*½ cup Sauce Vinaigrette**
2 cups leftover beef, chopped
*½ cup Mayonnaise**
2 tomatoes, cut into halves, sliced
10 radishes, cut into roses

Combine cauliflower, celery, olives, and Sauce Vinaigrette. Pile in the middle of a serving platter.

Combine the meat with Mayonnaise. Arrange around the salad on the platter. Garnish with a circle of tomatoes. Decorate with radishes.

SERVES 10.

. . . .

Hungarian Cabbage Strudel

Apart from its use in desserts, strudel dough is very helpful in making tasty hors d'oeuvres with different stuffings. Buy strudel dough in the dairy sections of supermarkets or gourmet shops specializing in imported foods. You can also use similar pastry for this dish—*filo*—available at Greek or Turkish food stores.

Cabbage strudel is perfect for a buffet with ham and roasts or meat balls. In Central Europe it is also often served for dinner as a first course on small plates together with clear meat or mushroom bouillon, clear tomato soup, and borsch.

Two sheets of strudel dough
¼ cup butter, melted

STUFFING
2 pounds head cabbage, grated
2 tablespoons salt
4 tablespoons bacon drippings
3 tablespoons sugar
¾ teaspoon pepper

TO MAKE STUFFING: Mix the cabbage with salt and let stand for 1 hour. Then squeeze out all the moisture. Heat the bacon drippings with the sugar, add cabbage and brown, stirring. Cover, and simmer over a very low heat until tender. Season with pepper and cool.

TO MAKE STRUDEL: Spread a damp large cloth napkin on a table. Unfold strudel dough. Place one sheet on the napkin. Brush with half of the butter. Place the second sheet over it. Brush with the remaining butter. Put stuffing along the short edge in a 4-inch strip. Roll with the aid of the napkin. Place on a buttered cookie sheet in 400° oven and bake for 30 minutes.

SERVES 8.

· ·

French Asparagus in Cheese Sauce

One of the nicest vegetables to serve at a party is asparagus. How fortunate we are to have it all year round in our freezers! But one can get tired of even the best vegetable if it is always served the same way. It pays to try to cook it as the French do—in a delicious sauce. This asparagus dish goes especially well with a roast, steak, or ham. Add tomato salad for color.

2 packages frozen asparagus
1 cup boiling water
1 teaspoon salt
½ teaspoon sugar
1 tablespoon butter

SAUCE
1 cup coffee cream
1 tablespoon cornstarch
2 egg yolks
½ cup grated Parmesan cheese
½ teaspoon nutmeg
Salt

IN THE AFTERNOON: Place asparagus in boiling water; add salt, sugar, and butter. Bring to a boil. Simmer covered for 5 minutes. Drain, saving the liquid. Place in a baking dish.

TO MAKE THE SAUCE: Heat the cream. Mix the cornstarch with egg yolks. Stir in the hot cream gradually. Bring to a boil, stirring constantly. Stir in half of the asparagus liquid; add ¾ of the cheese; bring to a boil. Season. Pour the sauce over the asparagus. Sprinkle with the rest of the cheese.

BEFORE SERVING: Place on a low shelf of the oven under the broiler for about 10 minutes or until golden.

SERVES 6.

. . . .

Baked Sole the Norwegian Way

No one in Europe knows how to cook fish better than the Scandinavians. The unyielding soil and the severe climate prevent good crops of vegetables and fruits. Swedes and Norwegians depend greatly on the sea and imported foods.

Baked fish is tastier and healthier than fried, also less time-consuming in its preparation. Use an attractrive baking dish and serve your fish right in it.

Serve with boiled potatoes or rice and a bowl of green salad. Well-chilled white table wine will supplement this dinner nicely.

> *3 pounds sole fillets*
> *6 tomatoes, fresh or canned, sliced*
> *Salt and pepper*
> *1½ cups whipping cream*
> *1 tablespoon green parsley, chopped*

Rinse sole fillets in a colander. Place in a buttered baking dish, 9″ × 12″. Cover with tomatoes. Season. Pour the cream

over tomatoes. Bake for 30 minutes in 360° oven. Sprinkle
with parsley before serving.

SERVES 8.

· ·

Turkey with Almond Stuffing from Prague

Women in Prague have an exceptionally neat appearance. Once, riding in a streetcar there, I noticed with surprise
that many ladies were standing, although there were plenty of
empty seats. After asking my Czech companion for an explana-
tion, I learned that they would rather stand than risk wrinkling
their clothes. Later I also observed with pleasure that they
cook as well as they dress.

> One 14–16-pound turkey
> ¼ cup butter, melted
> Salt
>
> STUFFING
> ½ pound French bread, sliced
> 1 cup milk
> 3 eggs, separated
> 2 tablespoons green parsley, chopped
> ¼ pound butter, melted
> ¼ teaspoon salt
> Rind grated from 1 lemon
> 2 ounces almonds, peeled, ground

TO MAKE STUFFING: Soak the bread in milk and squeeze
well. Combine with egg yolks. Add parsley; sauté in butter.

Beat egg whites with salt until stiff. Combine with lemon rind and almonds. Fold into the bread mixture.

TO ROAST TURKEY: Stuff the neck and body cavities; close with skewers. Brush with butter. Place in oven preheated to 450°. Set the thermostat to 325° and roast, counting 18 minutes per pound. Baste frequently with drippings. Sprinkle with salt when half done.

SERVES 12.

Turkey with Raisins from Vienna

During the last World War, an American spy was sent behind the front lines on a special mission. His knowledge of German was perfect; every detail of his behavior was prepared with meticulous care. One small omission, however, caused his real nationality to be discovered in a restaurant in Vienna while he was consuming turkey with gusto. The poor man was never taught to eat in the Continental way. Europeans never cut their meat into small pieces all at once, nor do they shift the fork to the right hand after every cut. They keep the knife all the time in the right hand, and eat with the left. Whoever does it in a different way may be immediately spotted as someone from overseas.

One 14–16-pound turkey
½ cup butter, melted
Salt

STUFFING
6 *slices white bread*
¾ cup milk
Turkey liver, finely chopped
4 tablespoons butter, melted
3 eggs, separated
½ cup raisins
3 tablespoons almonds, peeled, slivered
¼ teaspoon cloves
¼ teaspoon sugar
Dash allspice
⅓ teaspoon salt
1 cup bread crumbs

TO MAKE STUFFING: Soak the bread in milk and squeeze well. Combine with liver, butter, egg yolks, raisins, almonds, and spices. Beat the egg whites with salt until stiff. Add to the stuffing alternately with bread crumbs. Mix lightly.

TO ROAST TURKEY: See Turkey with Almond Stuffing.*

SERVES 12.

Baked Ham Continental Style

Baked ham is the simplest and comparatively the most inexpensive way of serving good meat to a large group. The Hawaiians have a nice way of baking ham with pineapple and heavy pineapple syrup. This, however, most Europeans find too sweet. They'd rather add tart apples to their pork and poultry.

Choose a boneless ham for easy slicing.

4–7 pounds fully cooked ham
½ cup brown sugar
¾ cup apple juice
2 teaspoons ground cloves

Place the ham in the roasting pan fat side up. Score the fat in squares. Roast in 325° oven for 1 hour.

Heat the sugar with apple juice and cloves. Stir till dissolved. Pour over the ham. Roast for another hour, basting with pan juices twice. Turn off the heat and leave the ham in the oven with the door half open for 15 minutes. Slice thinly with an electric knife.

SERVES 4 PER POUND.

. . . .

Spaghetti with Sausage the Bulgarian Way

My best friend at the University of Paris, Katia, a pretty brunette, loved to tell me about the life in her native Bulgaria. I never went to Sofia, but she described it so vividly that I could have sworn that I saw the streets covered with snow in the red glow of flickering flames while a Christmas Eve procession of people, each holding a burning candle, moved toward numerous domed, orthodox churches. But even Christmas delicacies never appealed to Katia as much as the simple dish of spaghetti and sausage her mother prepared so well.

1 pound spaghetti, cooked
1 pound sausage, sliced
2 tablespoons green parsley, chopped

SAUCE
2 carrots
1 parsnip
1 celery stalk
2 cups water
2 large onions, sliced
1 tablespoon bacon drippings
1 teaspoon paprika
2 tablespoons instant flour
1 cup tomato purée
Salt and pepper
¼ teaspoon sugar

TO MAKE SAUCE: Cook the carrots, parsnip, and celery in water for 20 minutes. Fry the onions in hot fat till golden. Stir in paprika and flour; fry for 2 minutes. Stir in the vegetable broth and tomato purée. Bring to a boil; add vegetables. Whip everything in a blender. Season. Pour the sauce over the sausage, simmer for 5 minutes.

Arrange spaghetti in a circle in a serving dish. Pour the sauce in the hollow. Sprinkle with parsley and serve.

SERVES 6.

English Roast Beef

In the opinion of some travelers, English cooking is not the greatest in the world. However, everyone must admit that there is nothing like the traditional English Sunday roast. It really does not have to be the most expensive cut for a tender and juicy result. The trick lies in low heat and proper timing. An English roast makes a simple but excellent dinner or a practical buffet dish that is as good when served hot as it is on a platter with cold cuts.

2½- to 5-pound eye round roast
Juice of 1 lemon
Salt and pepper

Sprinkle the roast with lemon juice, cover, and refrigerate for 24 hours, turning once or twice. Wash the roast and pound well. Place in a roasting pan fat side up. If it is very lean, sprinkle with melted butter. Cook at 300° for 45 minutes per pound if the roast is smaller than 4 pounds, 40 minutes per pound for larger cuts. Cut very thin slices, preferably with an electric knife; sprinkle with pan juices and serve.

ONE 3-POUND ROAST SERVES 8.

Hungarian Goulash

With the growing American interest in European cooking, Hungarian dishes have won a certain fame in the United States. The best known of all is Hungarian Goulash. Unfortunately, although the name appears on the menus of many restaurants, often just an inferior stew is served, having nothing to do with the real Hungarian national dish. The recipe below comes from a family of Hungarian refugees who fled their native land after the uprising of 1956.

2 medium onions, chopped finely
3 tablespoons lard or shortening
1 tablespoon paprika
2 pounds stewing beef chunks, trimmed and cut into
    1-inch cubes
Salt to taste
3 green peppers, sliced
2 tomatoes, cut into wedges
½ teaspoon caraway seeds
2 potatoes, cut into cubes
1 teaspoon tomato purée
Dash of garlic

Fry the onions in hot fat until golden; add paprika.

Sprinkle the meat with salt, add to the skillet, cover and simmer for 30 minutes. Add half of the green peppers and tomatoes, and the caraway seeds. Add a little water to make a thick sauce. Cover and simmer for 1 hour. Add the remaining green peppers and tomatoes, and the potatoes. Add some water if needed. Cover and simmer for another 30 minutes. Season with salt, tomato purée, and garlic. Serve with noodles or dumplings cooked separately.

SERVES 6.

Marjoram Goulash from the Great Plain

On the Great Plain of Hungary, herds of superior cattle graze. They have been there for many centuries, perhaps since the times when the nomadic Magyars came from the Urals to settle. Some variations of Hungarian goulash, or *Gulyas*, hark back to the times when it was stewed with vegetables in big kettles over campfires by shepherds moving from pasture to pasture.

Serve marjoram goulash with egg noodles and a green salad of lettuce and thinly sliced green peppers.

½ pound bacon, cut into long thin strips
4 onions, chopped
2 pounds round steak, cut into long thin strips
½ teaspoon marjoram
⅛ teaspoon pepper
Salt to taste
¾ cup white table wine
1 cup sour cream

Fry the bacon in a large skillet until half done. Remove and save. Add the onions to the bacon drippings; fry until golden brown. Spoon them out and save. Fry the meat in four portions. Combine the meat with onions. Season with marjoram, pepper and salt. Add wine, cover, and simmer for ½ hour. Add the bacon strips and simmer another ½ hour. Take off the heat, add sour cream, and serve.

SERVES 6.

. . . .

Rice Casserole Valencia Style

Spaniards come to life after dark. This is probably due to a scorching midday sun that makes everyone feel like taking a nice long *siesta* in the cool indoors. The Spanish nights are mild and fragrant with a thousand flowers, just perfect for a garden *fiesta,* which we Americans could call an evening cook-out.

Many of the delicious Spanish dishes are cooked on backyard grills. This one, called *paella*, is one of them, but it may be equally tasty prepared in the kitchen. The great variety of ingredients make you think of a New England clam bake.

Serve with lettuce and chilled rosé wine.

2 ½ pound chicken—breasts and legs
1 tablespoon flour
½ teaspoon salt
½ cup cooking oil
1 ½ pound large shrimp, shelled
1 cup kolbassi, sliced
1 medium onion, chopped
2 cloves garlic, crushed
1 pimento
1 cup canned tomatoes, chopped
2 cups uncooked rice
1 can minced clams and juice
½ package frozen peas
Salt to taste

BROTH
Remaining parts of chicken
5 cups water
1 ½ teaspoon salt
1 onion, quartered
½ teaspoon saffron

IN THE MORNING. Cook the broth: Cover chicken parts with water, add salt, onion, and saffron. Boil for 45 minutes. Strain.

Cut chicken breasts and legs into small pieces. Dust with flour and salt. Fry in hot oil until brown and tender. Remove from the skillet and reserve.

Fry shrimp and kolbassi until lightly browned; save.

ADD TO THE SKILLET: Onions, garlic, pimento, and tomatoes. Cook until the onions are tender. Refrigerate everything.

35 MINUTES BEFORE DINNER: Fry the rice in hot oil to glaze. Bring 4½ cups chicken broth to boil. Add the rice, tomato mixture, and clams, cook for 5 minutes. Add the peas, cook 5 more minutes uncovered. Add chicken, shrimp, kolbassi; mix, cover. Place in a 300° oven for 20 minutes.

SERVES 10.

- -

Transylvanian Lamb Casserole

In the long period of Turkish conquest in the 16th and 17th centuries, Hungarians tried very hard to preserve their national character, separating themselves almost totally from foreign influences. National traditions, costumes, folksongs, dances, decorative art, and cookery were carefully protected by Hungarian princes and gentry. However, in the province of Transylvania some favorite Turkish dishes found their way onto Hungarian tables.

2 pounds lamb, cut into 1-inch cubes
3 ½ cups water
Salt
⅛ teaspoon pepper
¼ teaspoon marjoram
¼ pound bacon, diced
1 large onion, chopped
1 tablespoon green parsley, chopped
1 ½ cups rice, uncooked (do not use instant rice)
½ cup grated cheese

Cover the meat with water; bring to a boil. Skim, season with salt, pepper, and marjoram; cover and simmer for 1 hour.

Fry the bacon with onions until golden. Add parsley and rice; fry for 5 more minutes. Add to the stewing lamb. Bring to a boil. Place covered in a 400° oven for 20 minutes.

Transfer to a warmed serving dish, sprinkle with cheese.

SERVES 8.

- -

Italian Pork Casserole

I have noticed many times at buffet parties that a beautifully arranged platter of cold cuts remained barely touched while dishes of homemade foods have been scraped clean. People are tired of catered meats and monotonous restaurant food. But they never tire of home cooking, especially when the hostess has some imagination.

Italian mothers excel in splendid casseroles, which are ideal for buffet entertaining, as they are easy to eat without using a knife. When the delicious aroma of this savory dish starts coming out of the kitchen during the baking time, you can almost hear a tarantella.

4 cups canned tomatoes
Salt and pepper
¼ teaspoon basil
¼ teaspoon marjoram
½ teaspoon soda
1 pound noodles
6 slices bacon, diced
6 onions, chopped
1 ½ pound fresh, lean pork, cut into small pieces
½ pound Italian pizza cheese, grated
½ cup bread crumbs

IN THE MORNING: Cook the tomatoes for 10 minutes; season with salt, pepper, basil, marjoram, and soda. Cook the noodles in salted water for 10 minutes; drain.

Fry the bacon until golden, scoop out, and save. Add the onions to the skillet, fry until golden brown, remove, and save. Brown the pork; season with salt and pepper.

Mix the tomatoes, noodles, bacon, onions, and meat. Place in a 9" × 12" buttered baking dish. Sprinkle with cheese and bread crumbs, cover, refrigerate.

2½ HOURS BEFORE SERVING: Let the casserole stand at room temperature for 1 hour. Bake in a 250° oven for 1½ hours.

SERVES 12.

- -

Pork and Cabbage from Helsinki

The custom of serving food straight from the oven to the table or buffet without special serving platters came to us from Scandinavia. It has the great convenience of keeping the dish warm until everyone is served and even ready for second helpings. The lovely colorful Scandinavian baking dishes and enameled pans are especially pleasant for this purpose.

2 pounds pork shoulder, cut into small pieces
2 tablespoons fat
1 small head white cabbage, shredded
1 cup frozen sliced carrots
½ teaspoon pepper
Salt to taste
Boiling water
2 tablespoons chopped green parsley

Brown pork in hot fat in a heavy kettle. Remove the meat. Arrange cabbage, carrots, and pork in layers, sprinkling each layer with salt and pepper. Pour in boiling water to barely cover. Simmer tightly covered for 2 hours. Sprinkle with parsley and serve.

SERVES 8.

Hungarian Sauerkraut Goulash

Hungarians are masters at stewing meats. They produce endless variations of economical, tasty one-pot dishes, especially suitable for large buffet parties. The following goulash is one of their best.

½ pound bacon, diced
4 onions, sliced
2 tablespoons paprika
1 tablespoon caraway seeds
⅛ teaspoon garlic
1 cup water
4 pounds stewing beef, cut into 1-ounce cubes
2 green peppers, sliced
Salt
2 pounds sauerkraut, drained, rinsed
2 cups cooked rice
1 pint sour cream

PREPARE A DAY AHEAD: Fry the bacon until transparent. Add the onions, fry until golden. Add paprika, caraway seeds, and garlic, and mix.

Transfer to a kettle, add water, meat, and peppers; season with salt. Simmer covered for 1 hour.

Add sauerkraut; simmer for another hour. Refrigerate.

BEFORE SERVING: Add rice, bring to a boil, stirring. Take off the heat, add sour cream, and mix well.

SERVES 12.

. . . .

Lithuanian Hunters' Stew

The deep Lithuanian forests have always provided ideal hunting grounds. Hunting expeditions used to take along barrels of sauerkraut and large rings of pork sausage. While the gentlemen were busy pursuing the deer, the boar, or the bear, sauerkraut with sausage was simmered for hours in a big kettle hanging over an open fire in the woods. Later the hunters added to the stew whatever they found—wild birds, venison, mushrooms, berries, and wild fruits. Nothing could be more delicious than this beautifully simmered stew when darkness brought the tired men back to sit around the fire, drink, eat, and chat about the excitements of the day. The contents of the kettle were reheated each evening, and each time they tasted even better, because Hunters' Stew has to be reheated to achieve the proper flavor.

There are many varieties of Hunters' Stew. In Poland tomatoes are usually added.

4 pounds sauerkraut
2 cups bouillon from cubes
1 pound bacon, diced
2 large onions, sliced
2 cups kolbassi, diced
2 cups leftover roast or poultry, diced
    (do not use meatloaf or lamb)
1 pound mushrooms, quartered
8 prunes, pitted
½ teaspoon pepper
4 bay leaves
Salt to taste
1 cup red, dry wine

Rinse the sauerkraut in a colander. Place in a kettle; add bouillon and ¾ of the bacon. Bring to a boil, cover, and simmer for 1 hour. Fry the onions with the rest of the bacon till golden. Add to the stew. Add kolbassi, meat, mushrooms, prunes, pepper, bay leaves, and salt. Simmer 1 hour longer. Refrigerate for 24 hours.

Before serving, add the wine, bring to a boil, cook on medium heat, stirring for 10 minutes. Remove the bay leaves.

SERVES 12.

Evening Snacks

PREFACE

My neighbor once told me that a new lady had joined her bridge group. "I don't like her," she said. "What's wrong, is she unpleasant?" I asked.

"Oh, no, but she's different."

There are still people in our country who wish to see only the expected things and to associate only with people who behave according to a certain predictable pattern. They reject whatever is different as inferior, or at best they just disregard it.

At the end of the 19th century, when huge waves of immigrants came to North America from every possible corner of the world, the newcomers clung together, finding security in the continuation of their national customs. The old-timers also clung together to protect their known way of life, fearing inundation by these foreign masses. To protect their position, the oldtimers placed their own traditions above all others. In that way English toast gained a superior status to Hungarian rye bread, English marmalade was considered more refined than Czech plum jam, and British kippers, roasts, and pie were all regarded as high class, compared with Polish fruit soup, Jewish pancakes, or Italian veal scallopini.

To a young modern American, this approach is simply ridiculous and belongs to history. The highest walls of discrimination could not prevent all those more recent settlers and their children and grandchildren from becoming true Americans.

Certainly they had a lot to learn and a lot to offer. The inevitable fusion produced a unique culture—the richest in the world—not just British any more, but a true American culture, drawing freely from everything fine the newcomers brought with them to the new land.

In time even the most conservative could not deny that the Pennsylvania Dutch had something interesting in the pot. And soon curiosity led to the investigation of Italian spaghetti sauces, Swedish smorgasbord, Russian salads, and so on.

Today international cooking is not only completely accepted in America, but very much in fashion. You can find foods from around the world in every corner store; canned dishes featuring foreign foods are advertised on TV by individuals carefully chosen for their foreign accents; people want to learn more about making Continental dishes.

But while the culinary art of various countries has evoked great interest in the United States, it has not always evoked interest in the people of these nations. Too often it is just taken for granted that persons with foreign accents belong to different circles, and they are not invited to the homes of the old-timers.

Newcomers, however, can enrich your social gatherings, liven up the conversation, introduce some new ways of entertaining, and encourage you to try their native dishes. Once you have ventured on this road of getting acquainted with people and things which are "different," never again will you want to turn back to the boredom of the parties of your past, predictable to the slightest detail.

An evening get-together for a lively exchange of ideas ending with an informal, easy, light snack is a perfect way to entertain new friends and old. In this chapter you will find simple little dishes ideal for this purpose. They can be prepared well in advance and served straight from the refrigerator or reheated in no time. If you cook them in the morning or a day before, you can have a carefree evening, knowing that after the theater, a ball game, bowling, or skiing, you will have a delicious, original, and elegant treat for your friends.

These evening snacks are not for cocktail party nibbling. They provide a light, but definite meal. They can be used equally well for luncheons and for a buffet. They are very convenient to serve after a bridge game, a committee meeting, or for a Sunday night supper when you don't feel like cooking, but you want to offer something nice to your house guests, and you are glad that you thought about it before they arrived. In addition to the usual coffee you have only to remember to offer caffeine-free coffee or tea because many people, even young and healthy, prefer it at a late hour.

A tea bag dunked in a cup is never elegant, and a wet, dripping tea bag is never an appetizing view on the table. There are two acceptable ways of serving tea: the English and the Continental way. It is customary in England to place the tea bags in a tea pot filled with very hot water that was carefully measured for a proper strength. In central Europe people use a small (3 cup), heat-resistant tea pot that was rinsed with boiling water. They place several tea bags in it, pour in boiling water, then put the pot on a warm stove to keep it very hot

290 · CONTINENTAL ENTERTAINING FOR THE YOUNG AND BUSY

but not boiling until the strong essence of tea is just right. This essence is later mixed with hot water in each cup separately so that each person can have his tea as strong as he wishes.

Most of the following little dishes are extremely economical. They are a perfect solution to the problem of what to do with leftover fish and meats. I have chosen the recipes for this chapter mainly because they require little or no last-minute preparation, and at the same time they provide a pleasant diversion from our usual menu. Once you have tried them, you will be glad that you welcomed these foreign newcomers into your home.

. . . .

For Evening Snack Menus

LIGHT DISHES

1. Stuffed Tomatoes from Andalusia
2. French Tomatoes with Chicken
3. Stuffed Tomatoes from the Netherlands
4. Tomatoes Stuffed with Shrimp from the Vatican
5. Stuffed Eggs from the Island of Ré
6. Eggs Stuffed with Crab from Russia
7. Shrimp from Finland
8. Stuffed Cucumbers from Stuttgart
9. Parisian Fish in Mayonnaise
10. Salmon in Mayonnaise from France
11. Salmon from the Mediterranean
12. Flounder from Brussels
13. Crab au Gratin from Brussels
14. Scallops from Marseilles
15. Norwegian Open Sandwich Smorgasbord
16. Danish Turkey Sandwich
17. Danish Steak Tartare Sandwich
18. Scandinavian Dry Fruit Sandwich
19. Banana Sandwich from Copenhagen
20. Italian Busy Girl Pizza
21. Cheese Fondue from Switzerland
22. Cheese Fondue from Lausanne
23. Meat Fondue from French Switzerland
24. Egg Fondue from Alsace
25. Mushrooms from the Tatra Region
26. Mushroom Pie from Paris
27. Mushroom Pancakes from Belgium
28. Polish Pancakes with Meat
29. Hungarian Pancakes
30. Omelet with Rum from Vienna

COOKIES

31. *Chocolate Squares from Mazury*
32. *Fig Squares from Poland*
33. *Apple Turnovers from Zürich*
34. *Turkish Diamonds*
35. *Rum Balls from Brno*
36. *Nutty Squares from Yugoslavia*
37. *Viennese Hazelnuts*
38. *Hungarian Almond Cookies*
39. *Almond Cookies from Dubrovnik*
40. *Polish Crullers*
41. *Norwegian Crullers*
42. *Greek Baklava*
43. *Swedish Ginger Hearts*
44. *Christmas Stars from Finland*
45. *Almond Crescents from Rumania*
46. *Walnut Cookies from Prague*

. . . .

Stuffed Tomatoes from Andalusia

The warm southern part of Spain is inhabited by a gay people who believe their region is the most beautiful in the world. And they are not just bragging. Andalusia is a beautiful spot, with its high mountains, fertile fields, and lovely sea coast. It is also famous for its excellent wines. But it is not only the vineyards that are handsome in Andalusia. The men are tall, and the women are known for their good looks. They wear colorful costumes, love to dance, and know how to cook, particularly light dishes appropriate for hot summer nights.

Serve these stuffed tomatoes with cold slices of good lean ham and warm garlic bread. Spanish or Portuguese rosé wine, well chilled, will make the snack perfect.

2 cups rice, cooked
½ cup Mayonnaise*
1 teaspoon prepared mustard
1 green pepper, seeds removed, chopped
2 tablespoons onions, finely chopped
3 tablespoons salad oil
6 medium tomatoes
Green parsley sprigs

Cook rice with a little less water than the directions call for. Cool. Combine with Mayonnaise mixed with mustard.

Sauté green peppers and onions in hot oil for 15 minutes over medium heat. Combine with rice. Stuff tomatoes. Refrigerate under cover until serving. Garnish with parsley.

SERVES 6.

. . . .

French Tomatoes with Chicken

Here is an inexpensive treat that will not keep the hostess in the kitchen. The Marquise of Sévigné, who three hundred years ago described so well the life in France of her era, served these tomatoes to her guests. She did so not for economic reasons, but because she liked tomatoes stuffed with chicken so much. In France chicken used to be and still is a costly meat dish.

Serve with hot French croissants and domestic dry white wine.

4 ounces mushrooms, coarsely chopped
2 tablespoons salad oil
2 cups cooked chicken, cubed
¾ cup Mayonnaise*
6 small tomatoes
Salt and pepper
½ green pepper, sliced
Parsley sprigs

Sauté mushrooms in hot oil for 5 minutes at medium heat. Combine with chicken and ½ cup Mayonnaise.

Cut off the tops of tomatoes. Remove central pulp. Sprinkle the interior with salt and pepper. Stuff. Refrigerate under cover.

BEFORE SERVING: Spread the tops of the tomatoes with the remaining Mayonnaise. Garnish with green pepper. Decorate the platter with parsley.

SERVES 6.

Stuffed Tomatoes from the Netherlands

More than three hundred bridges cross the canals of Amsterdam. Small boats move along while cars drive on either side. The strong and healthy Dutch are hearty eaters and love food. Thick pea soup is a favorite dish. Vegetables are usually mashed together with potatoes. Hollanders are great fishermen, and fish is often served at family meals and in many hors d'oeuvres and snacks.

Serve this dish with hot rolls, sweet butter, and wine.

1 large cucumber, peeled, sliced
2 teaspoons salt
6 small tomatoes
3 slices smoked salmon, chopped
6 tablespoons Mayonnaise*
1 hard-boiled egg, cut into 6 slices

Sprinkle the cucumbers with salt; refrigerate for ½ hour.

Cut off the tops of the tomatoes, remove half of the central pulp. Drain. Combine salmon with Mayonnaise. Stuff tomatoes. Place one slice of egg on the top of each tomato. Arrange tomatoes on a serving platter. Drain the cucumbers and place around the tomatoes.

SERVES 6.

. . . .

Tomatoes Stuffed with Shrimp from the Vatican

Each year thousands of visitors come to the residence of the head of the Catholic Church. There are many good restaurants in the vicinity of the Vatican which try to accommodate the crowds. This dish is often served and pleases people from different parts of the world; it is also a favorite with Americans.

Serve with Italian bread, plenty of butter, and Chianti.

6 tomatoes, medium size
Salt
36 medium shrimp, cooked
6 tablespoons Mayonnaise*
Parsley sprigs

Cut off the tops of the tomatoes and reserve. Remove half of the central pulp. Sprinkle the interior with salt. Place 5 shrimp in each tomato. Cover with Mayonnaise. Place tomato tops over it. Cut a little hole and stick a shrimp through each tomato top. Arrange on a platter; garnish with green parsley.

SERVES 6.

. . . .

Stuffed Eggs from the Island of Ré

A little French island on the Atlantic coast, Ré is a peaceful Eden with small coves and golden beaches. The local inns take full advantage of plentiful fish and shellfish from the ocean, preparing them in many ways for the gourmet's delight.

When you are planning a small party, serve fish for the family on the previous day. Cook more than you need, and use the leftovers to stuff some eggs.

Never boil eggs taken straight from the refrigerator. To prevent breakage, place them in warm water for 5 minutes.

1 cup cooked fish
6 hard-boiled eggs, cut into halves
2 tablespoons butter, soft
1 tablespoon paprika
Salt
Tomato Salad*
Green parsley sprigs

Press the fish and egg yolks through a strainer. Combine with butter, season with paprika and salt. Stuff the egg whites.

Arrange stuffed eggs on a serving platter around tomato salad. Garnish with parsley.

SERVES 6.

. . . .

Eggs Stuffed with Crab from Russia

This is an elegant dish for special company. It can be prepared well in advance and stay fresh until you return from the show with your guests or until your card game is over.

Arrange the stuffed eggs around a heap of your favorite vegetable salad. Serve with warm French croissants and white wine.

6 hard-boiled eggs, cut into halves
6 ounces canned crab meat, chopped
¾ cup Mayonnaise*
Lettuce leaves, shredded

Remove the egg yolks, and press them through a strainer. Combine the crab meat with half the egg yolks and ½ cup Mayonnaise. Stuff egg whites with the mixture. Spread each stuffed egg with the remaining Mayonnaise. Sprinkle with the remaining egg yolk.

Arrange stuffed eggs on a bed of lettuce.

SERVES 6.

Shrimp from Finland

Throughout northern Europe, crayfish is considered a special delicacy. It is caught in inland streams and sold alive in big baskets at farmers' markets for use in salads and soups, or served as a ragout on special occasions.

As crayfish are not easily found in American stores, we can substitute jumbo shrimp.

Prepare the dressing in advance. Serve with hot French bread and sweet butter.

1 head lettuce, well chilled
30 jumbo shrimp, cooked
6 hard-boiled eggs, sliced

DILL DRESSING
⅓ cup prepared yellow mustard
2 teaspoons sugar
⅓ cup wine vinegar
1 cup salad oil
2 tablespoons dill weed
Salt
White pepper

Line the salad bowl with lettuce leaves. Shred the remaining lettuce. Arrange lettuce, shrimp, and egg slices in layers in the bowl.

Mix the ingredients of the dressing, pour over the salad. Refrigerate for 10 minutes.

SERVES 6.

. . . .

Stuffed Cucumbers from Stuttgart

Here is a tasty snack combining French elegance with the German gusto for sour cream. It is simple to prepare, and a good way to use up leftover ham.

Serve with warm French bread and white Rhine wine.

2 cucumbers, peeled
1 teaspoon salt
⅛ teaspoon pepper
1 tablespoon vinegar
2 cups lean ham, finely chopped
¾ cup sour cream
¼ cup prepared horseradish

Cut each cucumber into six thick slices. Remove the seeds. Sprinkle with salt, pepper, and vinegar. Let stand for 2 hours. Drain. Combine ham with sour cream. Arrange cucumbers on a serving platter. Stuff the hollows with ham. Spread the stuffing with horseradish.

SERVES 6.

. . . .

. . . .

Parisian Fish in Mayonnaise

Parisians, who love night life so much, excel at creating small, light dishes that are perfect for midnight snacks. If you are planning to serve fish in Mayonnaise after a night out or a party, it is practical to have fish for dinner on the previous day, cooking an extra pound and saving it. It may be baked, broiled, or boiled. Fillets of sole, flounder, white fish, or cod will do nicely. For variety, canned salmon may be used.

Serve with warm French rolls or French bread, and sweet butter. White French wine will lend elegance to the dish.

1 package frozen baby lima beans
2 hard-boiled eggs, sliced
1 pound fish, cooked
6 lettuce leaves
3 tomatoes, sliced
4 ounces black olives
¾ cup Mayonnaise*

SAUCE FOR THE BEANS
1 tablespoon Mayonnaise
1 tablespoon salad oil
½ teaspoon prepared mustard
½ teaspoon vinegar
Salt to taste

IN THE AFTERNOON: Cook, drain, and cool the lima beans. Mix with the sauce, refrigerate. Prepare the eggs. Cut the fish into 1-inch pieces.

BEFORE SERVING: Place lettuce leaves on 6 medium plates. Arrange the lima beans and fish on each plate in an attractive way. Garnish with tomato and egg slices and olives. Cover the fish and dot eggs and tomatoes with Mayonnaise.

SERVES 6.

. . . .

Salmon in Mayonnaise from France

This little snack requires no cooking. Arrange the food on each separate small plate in a similar and attractive way. Serve with hot croissants, sweet butter, and rosé wine.

6 lettuce leaves
1 pound canned salmon, coarsely chopped
½ cup Mayonnaise*
3 hard-boiled eggs, sliced
12 slices dill pickle
6 fillets of anchovy
½ cup lettuce, finely shredded

Line 6 plates with lettuce leaves. Place a portion of salmon on each plate. Spread with Mayonnaise; cover with egg slices. Decorate the top with pickle slices and anchovy fillets. Sprinkle with lettuce.

SERVES 6.

. . . .

Salmon from the Mediterranean

Serving a combination of seafood has become very popular. This one is an attractive change from our customary fried fish and shrimp plate.

Serve with warm French bread and white wine.

6 small salmon steaks
1 celery stalk
1 onion, quartered
1 parsnip
Salt to taste
12 medium shrimp, cooked
12 rolled fillets of anchovy
3 canned beets, sliced
1 cucumber, peeled, sliced
1 tablespoon lemon juice
1 cup Mayonnaise*

Place fish steaks, celery, onion, and parsnip in a saucepan. Cover with boiling water; season with salt. Simmer for 20 minutes. Let stand in the water until cold.

Place fish steaks on a long serving platter. Place one shrimp and one fillet of anchovy on each side of each steak. Arrange cucumber and beet slices along the sides of the platter. Sprinkle salmon and vegetables with lemon juice. Garnish with Mayonnaise.

SERVES 6.

. . . .

Flounder from Brussels

Food was scarce in Europe during World War II and the first months after the Hitler's defeat. Belgium, liberated by the Allies, was licking her wounds and trying her best to entertain weary soldiers coming to Brussels for short leaves of rest. Wishing to live up to their culinary tradition in those difficult circumstances, the great restaurants took full advantage of the abundant fresh sea fish, serving it in the most imaginative ways in order to avoid monotony.

2 pounds flounder fillets
½ pound butter
½ cup spinach leaves, finely chopped
¼ cup parsley, finely chopped
1 teaspoon sage
1 cup white dry wine
1 cup water
Salt and pepper
3 egg yolks
Juice of 1 lemon
1 teaspoon cornstarch

Place flounder fillets in hot butter; cover with herbs. Sauté for 5 minutes. Add wine, water, salt, and pepper. Cover and simmer for 20 minutes.

Combine egg yolks with lemon juice and cornstarch. Stir in the liquid from the fish by spoonfuls until the cup is full. Add to the sauce in the pan. Heat everything, stirring gently, but do not boil. Transfer fish to a serving platter. Cover with the sauce. Serve warm or cold.

SERVES 6.

Crab au Gratin from Brussels

When we went to Brussels a few months after World War II, we visited a Jewish jeweler, bringing him a message from his brother in Germany. As the postal service between these countries was not established yet, this was the first time the jeweler heard that his brother had survived Hitler's massacre. Delighted, he treated us to a night in the best restaurants and clubs of the city.

I ended this fascinating evening with crab au gratin. Serve with lettuce and white wine.

One 4-ounce can crab meat
4 ounces frozen shrimp
12 soda crackers
1 cup shredded cheese

SAUCE BÉCHAMEL
3 tablespoons butter
4 tablespoons instantized flour
1 ½ cups milk
1 tablespoon sherry
Salt to taste
Dash white pepper
1 teaspoon dill weed

TO MAKE SAUCE: Mix the butter with the flour over low heat. Gradually stir in the milk and sherry. Bring to a boil, stirring constantly. Season with salt, pepper, and dill.

Drain the crab meat, and dilute the sauce with the liquid. Combine crab meat and shrimp with the sauce.

Transfer to a well-buttered square shallow baking dish. Cover with crackers. Sprinkle with cheese. Place in a 400° oven for 20 minutes or until golden.

SERVES 4.

. . . .

Scallops from Marseilles

French cuisine is one of the most refined in the world. Simplicity is not its virtue, and many French recipes scare us away, as nowadays we seldom wish to spend the considerable time in the kitchen needed to prepare them. However, with proper selection and modern methods, we can cook the French way and still not overwork ourselves.

This light dish, excellent for late snacks or luncheons, is one of the easiest French recipes and always successful. Serve it with colorful tossed salad and French dressing, hot croissants, and well-chilled white table wine.

1 pound scallops, cut in quarters
1 pound mushrooms, sliced
5 tablespoons butter
3 tablespoons lemon juice
1 cup white table wine
1 bay leaf
½ teaspoon salt
⅛ teaspoon pepper
3 tablespoons instantized flour
1 cup coffee cream
⅔ cup bread crumbs

IN THE MORNING: Wash and cut the scallops. Fry the mushrooms in 2 tablespoons of butter until golden; add the lemon. Combine the wine, bay leaf, salt, and pepper in a saucepan. Add scallops. Cover and cook on low heat for 10 minutes.

Drain, save the broth. Remove the bay leaf. Melt the rest of the butter in a saucepan, stir in flour. Stir in 1 cup of the saved broth and cream. Cook over low heat, stirring until sauce is smooth and thickened.

Add scallop mixture and mushrooms. Mix well. Refrigerate.

BEFORE SERVING: Spoon into 6 sea shells or heat-resistant separate small dishes. (It may also be done in one baking dish.) Sprinkle with bread crumbs; dot with butter. Bake in a 400° oven for 10 minutes or until golden.

SERVES 6.

· ·

Norwegian Open Sandwich Smorgasbord

I'll bet you've never seen anything like this: one huge sandwich loaded with a variety of foods that will feed four people. But in Norway this hearty sandwich is often served after bridge or for teen-agers' parties.

Garnish with tomato wedges and lettuce leaves. Serve in the same dish it was baked. Cut on the table.

12-inch French bread cut lengthwise in two
¼ pound sweet butter
10 slices ham
2 cans asparagus tips, well drained
16 wieners
⅓ cup mustard
1 onion, sliced
2 eggs
1 cup coffee cream
4 cups grated cheese

Spread both halves of bread with butter. Cover with ham. Arrange asparagus and wieners in alternate rows. Spread the wieners with mustard. Cover everything with onions.

Beat the eggs with cream and cheese. Pour over sandwiches. Bake in hot 475° oven for about 10 minutes or until golden brown. Cut each sandwich into 4 portions.

SERVES 8.

. . . .

Danish Turkey Sandwich

Serve one or two kinds of nourishing, tasty, open sandwiches after an evening at the bridge tables. Prepare everything before your guests arrive. Cover tightly with plastic wrap and refrigerate. Coffee with cookies may follow.

8 slices pumpernickel
4 tablespoons sweet butter
1 head bib lettuce
1 cucumber, sliced unpeeled
16 thin slices turkey
16 slices jellied cranberry sauce
8 sprigs parsley

Spread the bread with butter. Cover half of each bread slice with lettuce, half with cucumber. Cover with turkey. Garnish with cranberry sauce and parsley.

SERVES 8.

. . . .

. . . .

Danish Steak Tartare Sandwich

Tivoli amusement park in Copenhagen glitters at dusk with a million lights. It is a giant Christmas illumination for every day, an electrical extravaganza made to lift the spirits of people from the North who are so often afflicted with depression during their long dark winters.

Not every tourist strolling among the hundreds of excellent restaurants in the Tivoli gardens can afford a full-course meal. Famous Danish open sandwiches provide an attractive consolation.

1 ½ pounds twice-ground lean sirloin
1 ½ teaspoons salt
⅓ teaspoon freshly ground pepper
¼ cup minced onions
6 slices pumpernickel
3 tablespoons butter
6 tablespoons prepared horseradish
1 cup sliced black olives
6 small lettuce leaves
6 small egg yolks, white threads carefully removed
6 half egg shells

Combine the meat with salt, pepper, and onions. Spread the bread with butter; cover with steak tartare. Place one tablespoon of horseradish in the center of each sandwich. Refrigerate for ½ hour.

Arrange olives on one side; place a lettuce leaf on the other side of each sandwich. Place each egg yolk in the empty egg shell on the lettuce. Serve immediately.

Everyone spreads the egg yolk over his own meat.

SERVES 6.

Scandinavian Dry Fruit Sandwich

This is a sandwich that is popular for hen parties. It is original and simple; your friends will love it. Prepare everything before they arrive and keep refrigerated.

16 pitted prunes
16 apricot halves
8 pear halves
8 slices pumpernickel
4 tablespoons sweet butter
1 head butter lettuce
8 thick slices of meatloaf
8 sprigs of parsley

Soak dry fruits in water for 24 hours. Drain well.
Spread the bread with butter. Cover with lettuce leaves and meatloaf. Garnish with dry fruits and parsley.

SERVES 8.

Banana Sandwich from Copenhagen

In Denmark sandwiches are always open and loaded with so many things that each one makes a meal in itself.

This is a typical Danish after-theater snack. You can prepare everything and refrigerate before going out. Place your sandwiches in the oven on a cookie sheet just before serving.

Serve with well-chilled beer in large tumblers. A basket of assorted fruits may follow for dessert.

6 slices French bread, lengthwise
3 tablespoons sweet butter, soft
6 slices lean ham
3 small bananas, sliced lengthwise
3 hard-boiled eggs, coarsely chopped
½ pound mushrooms, slices, sautéed in butter
6 slices Swiss cheese
6 small wedges tomatoes
6 small sprigs of parsley

Spread bread slices with butter. Cover with ham. Place banana slices diagonally in the middle. Place the egg on one side of bananas, mushrooms on the other. Cover with cheese.

Bake in 450° oven for about 10 minutes until golden brown. Garnish with tomato and parsley and serve immediately.

SERVES 6.

Italian Busy Girl Pizza

This is a quick treat for a teen party, much less expensive and more original than store-bought pizza—a tasty home-cooked dish that does not require mother's supervision. There's no messy kitchen, it takes just 10 minutes to assemble (before the crowd comes), then 15 minutes in the oven—and it's ready to serve.

1 *loaf French bread, cut horizontally and lengthwise*
  into 4 slices
¼ *pound margarine, soft*
½ *cup canned tomato purée*
1 *teaspoon Italian herb seasoning*
1 *teaspoon oregano*
½ *pound wieners or sausage, sliced*
1 *green pepper, seeds removed, sliced*
3 *large tomatoes, sliced*
¼ *pound pizza cheese, shredded*

Spread bread slices with margarine. Arrange on a buttered cookie sheet forming a rectangle. Combine tomato purée with herbs and spread on the bread. Arrange wiener slices and green pepper on it. Cover with tomato slices. Sprinkle with cheese.

Bake in a 400° oven for 15 minutes or until golden brown.

SERVES 8.

Cheese Fondue from Switzerland

Fondue has become very fashionable in our country. It is one of Switzerland's national dishes, very informal and usually prepared for the family and close friends. In Switzerland fondue is never served as an hors d'oeuvre, but as a meal in itself. Quite often on Sundays, people don't want to be bothered with an elaborate dinner. Cheese fondue is served instead. It is prepared in the kitchen in special earthenware dishes and brought to the living room to be consumed in front of the fireplace. Everyone is given a bowl with cubed white bread and a fork with a long handle. The fondue is kept warm over a small burner.

Etiquette requires one to eat the bread dipped in melted cheeses without the fork touching the lips. This, however, cannot be always avoided, and for that reason fondue is served only in small intimate groups. Young people especially like it on frosty evenings after skiing and other winter sports.

1 clove garlic
3 cups white dry wine
10 ounces Swiss cheese, grated
10 ounces Gruyère cheese, grated
10 ounces Fontina cheese, grated
3 tablespoons kirsch
1 ½ teaspoons cornstarch
Dash white pepper
Dash nutmeg
1 teaspoon butter
Narrow, crusty French bread, cut into 1-inch cubes

Rub the bottom of the fondue dish with garlic. Add wine. Warm on medium heat until foamy, but do not boil. Add the cheeses gradually, stirring. When melted, pour in the kirsch mixed with cornstarch. Season with pepper and nutmeg. Add butter. Continue to cook until the fondue begins to thicken.

SERVES 6.

. . . .

Cheese Fondue from Lausanne

Cheese fondue as served to the Swiss family or at small teenage parties in Lausanne, a French-speaking city in Switzerland, is simple to prepare. Cooked shrimp, diced ham, and pieces of wieners are often served for dipping in the fondue in addition to crusty cubed bread.

1 clove garlic
2 cups dry white wine
20 ounces Swiss or Gouda cheese, cut into small bits
2 tablespoons cherry cordial
Salt to taste
1 teaspoon cornstarch

Smear the bottom of the fondue dish with garlic. Add wine and warm up on medium heat until foamy. Add the cheese gradually, stirring. When melted, add cordial mixed with cornstarch. Season and continue to cook until the fondue begins to thicken. Serve immediately.

SERVES 4.

Meat Fondue from French Switzerland

The French-speaking Swiss often serve *Fondu Bourguignonne,* which is very much to the taste of American beef lovers. Choose a prime, tender piece of steak, the best you can afford. Two or three kinds of sauce must be served for the proper appreciation of this dish. Sauce Hollandaise* and Sauce Bordelaise* are recommended. For the third one serve your favorite barbeque sauce. A bowl of tossed salad and French bread will supplement the dish well.

2 ½ pounds steak, very well trimmed, cut into
small cubes
Fondue pot half filled with oil

Heat the oil in the fondue pot, bring to the living room and keep warm on a fondue burner. Serve meat cubes and sauces in separate bowls. Everyone cooks his own meat, dipping it on a long fork in the oil, then in one of the sauces.

SERVES 6.

Egg Fondue from Alsace

Alsace is neither truly German or truly French in character, but a mixture of customs from both these countries, and in addition it has its very own Continental charm and elegance. This fondue, different from the popular cheese one, is very nice at the end of a long evening gathering.

> 4 eggs, slightly beaten
> ¼ pound sweet butter in small bits
> ½ pound Swiss cheese, grated
> ½ cup dry white wine
> Salt and pepper to taste
> Cubed crusty French bread

Cook the eggs very slowly in a buttered fondue pan, do not allow them to set. Add butter, mix, add cheese. Stir in the wine, season, heat and serve promptly.

SERVES 4.

Mushrooms from the Tatra Region

The late summer and early fall is mushroom season in the Old World, which still depends to a large degree on crops gathered in the woods. In this country we now have mush-rooms all year round, but somehow when autumn approaches, mushroom dishes seem very much in harmony with nature.

Serve with hot rolls, sweet butter, assorted cheeses, and wine.

2 pounds mushrooms
Salt to taste
2 teaspoons caraway seeds
4 medium onions, finely chopped
6 tablespoons salad oil
¼ cup lemon juice
Salt and pepper
2 tablespoons green parsley, finely chopped
2 teaspoons dill weed
2 large tomatoes, cut into wedges
3 hard-boiled eggs, quartered
3 tablespoons Mayonnaise*

Trim the ends; rinse the mushrooms thoroughly. Place in a saucepan; cover with boiling water; add salt and caraway seeds. Cook vigorously for 5 minutes. Drain.

Slice large mushrooms and cut medium ones into halves. Combine with onions. Season with oil, lemon, salt, and pep-per. Refrigerate for 2 hours.

Arrange mushrooms into a heap on a serving platter. Sprinkle with parsley and dill. Garnish with a circle of eggs and tomatoes. Dot eggs and tomatoes with Mayonnaise.

SERVES 6.

Mushroom Pie from Paris

It is really a shame that we use so few mushrooms in our cooking. Europeans are very fond of them, and they know

many ways of serving them. Gathering wild mushrooms is a
favorite summer time occupation in many European countries.

Wild mushrooms have more aroma, but it is possible to
make this original, delicious mushroom pie even with those
we can buy in supermarkets. Serve it with tomato salad, white
wine, and warm French bread.

1 9-inch pie shell
2 pounds mushrooms
4 tablespoons butter
Salt and pepper
1 cup sour cream
1 tablespoon chopped green parsley

IN THE MORNING: Bake a pie shell from a mix. Wash
mushrooms in a colander. Cut off a small slice of the stems. Cut
into halves. Heat the butter in a saucepan, add the mushrooms, and cover. Cook slowly for 20 minutes. Refrigerate.

BEFORE SERVING: Heat the pie shell in the oven. Reheat
the mushrooms, stirring. Add sour cream. Season. Heat, but
do not boil. Transfer the mushrooms into the pie shell, sprinkle with parsley. Serve immediately.

SERVES 6.

Mushroom Pancakes from Belgium

I was married in Germany a few months after the end of
World War II, when it was still very difficult to buy anything.
To please me, my fiancé made a special trip to Belgium and
brought me a white veil for the wedding, a few bottles of
champagne, and a gift from his Belgian friend—a framed
favorite family recipe.

BATTER
1 cup milk
2 eggs
1 cup flour
½ cup water
¼ teaspoon salt
3 tablespoons butter

STUFFING
½ pound mushrooms
Juice of 1 lemon
6 ounces ham, chopped
¼ pound sweet butter
1 tablespoon instant flour
1 cup coffee cream
2 tablespoons grated cheese
Pinch pepper

IN THE AFTERNOON:
TO MAKE PANCAKES: Beat the milk with eggs and flour for 2 minutes. Add water and salt; beat some more. Heat a 6-inch skillet and brush with butter. Pour in 3 tablespoons batter and tilt the pan immediately so that the batter will spread over entire bottom. Cook the pancakes on medium heat on both sides. Stack on a plate.

TO MAKE STUFFING: Chop the mushrooms and sprinkle with lemon juice. Sauté the ham in 1 tablespoon butter. Remove it from the pan; add remaining butter and the mushrooms. Sauté at high heat for 3 minutes. Add the ham, mix, sprinkle with flour. Stir in cream by spoonfuls until the stuffing has a sour cream consistency. Save the remaining cream. Add 1 tablespoon cheese. Season. Stuff the pancakes, folding the edges envelope fashion. Arrange in a buttered baking dish, sprinkle with the remaining cream and cheese. Cover and refrigerate.

BEFORE SERVING: Place uncovered in a 400° oven for about 20 minutes or until golden. Baste twice.

SERVES 6.

. . . .

Polish Pancakes with Meat

Polish people love parties and food that goes well with vodka. This is an excellent snack to serve after the show when you would like to linger around the table for a while discussing the night's entertainment with your guests.

Make those pancakes dainty and elegant. Prepare them, stuff and refrigerate in the morning. Fry them just before serving. Serve very hot with a colorful tomato salad sprinkled with chopped green onions. Place a bottle of well-chilled Lemon Vodka* and small cordial glasses on the table.

BATTER
1 ½ cups milk
3 eggs
1 ½ cups flour
¾ cup water
½ teaspoon salt
2 tablespoons bacon drippings

MEAT STUFFING
1 large onion, chopped
2 tablespoons bacon drippings
3 cups leftover roast or poultry, ground
⅓ cup strong bouillon from a cube
Salt and pepper to taste

FOR FRYING
1 egg, beaten
¼ cup milk
½ cup bread crumbs
4 tablespoons bacon drippings

TO MAKE BATTER: Beat the milk with eggs and flour for 2 minutes. Add water and salt; beat some more.

TO MAKE PANCAKES: Heat a 6-inch skillet and brush with bacon drippings. Pour in 3 tablespoons batter, and tilt the pan immediately so that the batter will spread over entire bottom of the pan. Cook the pancake on medium heat on both sides. Repeat till all the batter is used. Stack the pancakes on a plate.

TO MAKE STUFFING: Fry the onions in bacon drippings until golden. Add meat, bouillon, salt and pepper. Mix well.

TO STUFF: Spoon a little meat stuffing into the center of each pancake and fold the edges envelope fashion to encase the stuffing completely. Refrigerate in a covered dish.

TO FRY: Roll each stuffed pancake in the egg mixed with milk and then in bread crumbs. Fry in hot fat until golden on both sides. Serve immediately.

SERVES 8.

. . . .

Hungarian Pancakes

Hungarian pancakes, called *palacsinta,* make a good late snack as they may be prepared well in advance and can easily be reheated. In Hungary they hold an important place among sweet dishes. This type of pancake is very popular all over Central Europe. The batter is similar everywhere, only the way they are stuffed, folded and reheated differs depending on the region.

BATTER
4 eggs, separated
2 cups milk
1 tablespoon sugar
2 cups flour
3 tablespoons butter, melted

SPREAD
1 cup pure apricot preserves
1 cup almonds, slivered
1 pint sour cream

Beat the egg yolks with milk, sugar, and flour. Whip the egg whites until stiff, fold into the batter.

Heat a 7-inch skillet. Brush lightly with butter. Pour in ¼ cup of the batter and tilt the pan immediately so that the batter will spread over entire bottom of the pan. Cook the pancake on both sides. Repeat until all the batter is used, stacking the pancakes on a plate.

Spread each pancake with preserves, and sprinkle with almonds, using ¾ of the almonds. Roll each pancake and place in a buttered baking dish. Spoon sour cream over the rolled pancakes; sprinkle with remaining almonds. Cover.

BEFORE SERVING: Place the covered dish with pancakes in 350° oven and bake for 15 minutes.

SERVES 8.

Omelet with Rum from Vienna

In all Viennese cookbooks more space is given to desserts than to everything else together. This original omelet may be served after a light dinner; however, it is principally used nowadays as an ending to a bridge party when drinks and various snacks have been served during the game. The preparation takes 5 minutes; the cooking time is 10 minutes.

8 eggs
1 teaspoon sugar
3 tablespoons butter
¼ cup confectioners' sugar
⅓ cup rum

Whip the eggs with sugar in a blender. Heat the butter in a 10-inch pan. Pour in the eggs. Fry over medium heat. Lift the edges of the omelet several times with a knife and allow some of the runny egg to flow underneath. When the top is still a little soft, transfer the omelet to a warmed serving platter. Sprinkle with confectioners' sugar through a sieve. Quickly heat the rum in the same pan, but do not allow to boil. Pour over the omelet, put it aflame, and immediately bring it to the table.

SERVES 6.

Chocolate Squares from Mazury

Frederic Chopin was born in a small village in the Polish province of Mazury. He loved his native land so much that he incorporated many folk motifs into his compositions. A number of them have the typical rhythm of the mazurka, the dance of this region that was very popular in 19th-century Europe. Chopin named his lively and sentimental masterpieces *"mazurkas"* after the dance. Polish people, who loved Chopin's music, called their famous thin cakes served at Easter time by the same name in his honor. Let's try this exquisite chocolate mazurka.

DOUGH
1½ sticks margarine, soft
2¾ cups sifted flour
⅔ cups confectioners' sugar
2 teaspoons baking powder
1 egg
1 egg yolk
4 tablespoons sour cream

SPREAD
12 ounces chocolate
⅓ cup coffee cream
4 large egg yolks
1 teaspoon vanilla extract
1¼ cups sugar
1 tablespoon instant flour
¼ cup almonds, slivered

TO MAKE DOUGH: Cut the margarine into the flour with a knife and rub in with finger tips. Add sugar and baking powder; mix. Add the remaining ingredients and knead the dough.

Roll out thin. Place on a buttered cookie sheet 12″ × 15″. Spread the dough until the sheet is almost covered. Bake at 375° for 20 minutes.

TO MAKE SPREAD: Melt the chocolate with cream in a warm oven; do not let it boil or dry out. Beat the egg yolks with vanilla for 3 minutes. Add sugar in small portions, still beating. Add the flour and beat 5 more minutes. Combine with the melted chocolate. Spread over the slightly cooled cake. Sprinkle with almonds. Bake in a 300° oven for 10 minutes. Cool. Cut into small squares and remove from the sheet with a knife.

YIELDS 48 SQUARES.

. . . .

Fig Squares from Poland

Thin Polish Easter cakes—mazurkas—come in endless varieties. The mazurka is usually a commercial white wafer or a very thin layer of cake and a generous layer of sweet spread made of eggs mixed with nuts, almonds, or various exotic fruits, or . . . everything together. Instead of figs you can use prunes or dates for a change.

BOTTOM LAYER
*Prepare the dough and bake
as for Chocolate Squares**

SPREAD
1 ¼ cups sugar
½ cup water
½ tablespoon vinegar
1 ½ cups walnuts, chopped
½ pound figs, ground
4 ounces candied orange rind, ground

324 · CONTINENTAL ENTERTAINING FOR THE YOUNG AND BUSY

ICING
2 cups confectioners' sugar
4 tablespoons lemon juice

TO MAKE SPREAD: Bring the sugar with water and vinegar to a boil. Cook for 6 minutes, stirring. Add the remaining ingredients. Cook for 12 minutes, stirring. Spread the warm filling over the cake. Cool.

TO MAKE ICING: Combine sugar with lemon juice. Spread over the mazurka. Refrigerate for 1 hour. Cut into small squares. Remove from the baking sheet with a knife.

YIELDS 4 DOZEN.

. . . .

Apple Turnovers from Zürich

Of all the fruits, apples are the most useful for baking. First, they are available all year round. Second, we will never tire of them, as there are hundreds of ways to serve them— in apple pies, cakes, and various pastries. Over the centuries people used their imagination to invent these desserts. In times when transportation was difficult and refrigeration unknown, in northern Europe apples were the only fruit one could eat during the long winter months. No other fruit could be preserved so well and for so long in cold cellars.

Tart apples are best for baking. Unfortunately they are sometimes difficult to get. Sprinkling them lightly with lemon juice will improve the taste of sweet apples.

This old recipe is an all-time favorite in the Alpine regions.

½ pound margarine
4 cups flour, sifted
½ ounce fresh yeast
1 teaspoon vanilla extract
1 cup sour cream
4 medium apples, peeled, cored, sliced
5 tablespoons sugar
½ cup confectioners' sugar

Cut the margarine into the flour with a knife and rub in with fingertips. Add yeast mixed with vanilla and sour cream. Knead the dough for a few minutes. Roll pastry into a square ⅛-inch thick or a little less. Cut into rectangles 3 × 2 inches. Sprinkle the apples with sugar. Place an apple slice on each rectangle. Lift two opposite ends, fold over the apple, press together. Lift turnovers carefully with a knife; place on a buttered cookie sheet. Bake in a 375° oven for 40 minutes or until golden.

Take the turnovers out of the oven and immediately sprinkle generously with confectioners' sugar.

YIELDS APPROXIMATELY 60 TURNOVERS.

. . . .

Turkish Diamonds

It often happens that we have to host a club meeting or entertain a few friends for bridge, and we wish to end the evening with something small, tasty and elegant, a little different from what everyone on the block serves. Here is your solution: an easy-to-prepare company dessert from the Near East. In Turkey it is a popular standard dish. But in this country you may be sure to be original.

As with most desserts from that part of Europe, it is

soaked in syrup, rather sweet, and should be cut into small portions.

Prepare it in the morning or on the previous night.

SYRUP
1 ¼ cups sugar
1 ½ cups water
2 teaspoons lemon juice

BATTER
3 eggs
1 cup sugar
1 cup plain yogurt
1 cup flour, sifted
1 teaspoon baking powder
1 teaspoon lemon rind, grated

DECORATION
½ cup whipped cream

Boil sugar with water for 5 minutes. Set aside to cool. Add lemon juice. Beat the eggs with sugar for 5 minutes. Add yogurt and flour mixed with baking powder and lemon rind. Blend well. Pour into a well-buttered 9-inch square baking pan. Bake in 400° oven for ½ hour. Cool.

Cut the pastry in the pan into diamonds. Pour the syrup over it. Let stand for several hours or until the syrup is absorbed.

Place each diamond on an individual plate. Dot with whipped cream.

SERVES 8.

. . . .

Rum Balls from Brno

Czechs are very thrifty. I really think that all the funny jokes about miserly Scotchmen could apply as well to Czechs. My landlady in Brno, for instance, once confided to me that she kept her wedding ring on her left hand (Europeans wear them on the right) because that way gold wears off less. Her white dog was lucky to get a bath once a week, but her black one had to wait a whole month.

The good woman had one weakness. When she made these rum balls, she couldn't resist them.

FILLING
¼ pound sweet butter, soft
½ cup confectioners' sugar
1 egg yolk
1 tablespoon rum
½ teaspoon strong coffee

DOUGH
1 cup walnuts, ground
1 egg white, slightly beaten
1 cup sugar

COATING
1 cup instant chocolate

TO MAKE FILLING: Cream the butter with sugar; add remaining ingredients; beat for 5 minutes. Refrigerate.

TO MAKE DOUGH: Combine the walnuts with egg white. Add enough sugar to make a good dough.

TO MAKE RUM BALLS: Cut out small pieces from the filling. Roll into balls. Wrap each ball in the nut dough. Roll in chocolate. Refrigerate.

Remove from the refrigerator for 1 hour before serving.

SERVES 8.

Nutty Squares from Yugoslavia

Some of the best old European recipes never appeared in cookbooks. They were jealously guarded as family treasures and handed down from mother to daughter.

During my travels in Yugoslavia, I stayed for a while in a small town with my friends, a local judge and his family. It happened to be Easter, and among many goodies I tasted these nutty cookies. Seeing my delight, my hostess gave me her recipe. She made that exception knowing that I would leave soon. I had to promise that I would not give it to any of her neighbors.

BATTER
11 large egg whites
2 cups confectioners' sugar, sifted
1 teaspoon vanilla extract
½ teaspoon almond extract
1 ½ cups walnuts, ground
1 ¼ cups almonds with peel, ground
½ cup instant chocolate powder

ICING
1 tablespoon butter
2 teaspoons instant coffee
3 tablespoons milk
1 ½ cups confectioners' sugar

TO MAKE THE BATTER: Beat the egg whites until stiff. Add the confectioners' sugar by spoonfuls, still beating. Add vanilla and almond extracts; beat for 10 minutes.

Mix the nuts with almonds and chocolate, fold into the egg whites, and mix lightly. Bake in a 9″ × 12″ pan, well buttered and sprinkled with bread crumbs, in a 325° oven for 45 minutes. Turn off the heat, and let the cake stand for 10 minutes with the oven door half open. Cool. The cake will fall a little.

TO MAKE ICING: Melt the butter with coffee and milk. Mix with sugar. Spread over the cake. Cool. Cut into small squares.

YIELDS 32 SQUARES.

- -

Viennese Hazelnuts

The most fabulous cakes I have ever seen and eaten in my life were invented in Vienna. French cuisine cannot even compare with the variety and refined tastes that came out of the romance the Viennese started with their bakers centuries ago.

⅔ cup sweet butter
1 teaspoon vanilla extract
3 tablespoons sugar
1 cup flour
1 cup ground hazelnuts or walnuts
1 ½ cups instant chocolate powder

Beat the butter with vanilla and sugar for 7 minutes. Add the flour; beat some more. Mix well with nuts.

Roll in your palms into marble-sized balls. Bake on a

buttered cookie sheet in 375° oven for 15 minutes. Immediately roll in chocolate.

YIELDS ABOUT 10 DOZEN.

. . . .

Hungarian Almond Cookies

Some time ago, according to a popular notion, Hungarian cuisine consisted mainly of goulash and noodles with cheese, and a few other things with impossible names, swimming in lard, smelling of onions, and burning tongues with its hot, red pepper. Today we know better. We know that Hungarians not only love to eat, but can cook with inspiration and produce a great variety of palatable dishes. They especially excel in home-baked cakes and dainty cookies.

It takes 15 minutes to prepare the dough and 40 minutes to bake these delicate almond cookies. Do it a day ahead of your party. They are better the next day, and you will not be tired.

½ pound butter, soft
1 ¼ cups sugar
8 ounces almonds, peeled, ground
1 teaspoon vanilla extract
1 tablespoon grated orange rind
5 egg whites
1 ½ cups flour

Beat the butter with the sugar for 5 minutes. Add the almonds, vanilla, and orange rind. Mix with a spoon.

Beat the egg whites until stiff. Add the flour and egg whites to the butter alternately in small portions. Mix everything together lightly. Spread over a buttered 9″ × 12″ pan. Bake in 350° for 40 minutes. Cool. Cut in the pan into small squares. Remove with a knife.

YIELDS 48 COOKIES.

· ·

Almond Cookies from Dubrovnik

Within the walled city of Dubrovnik in Yugoslavia, where close to 6,000 people live, no motor vehicles are permitted. You can stroll peacefully for hours admiring ancient narrow streets, or stop at one of numerous cafés and sip coffee, crunching cookies made from homegrown almonds.

½ pound sweet butter, soft
1 ¾ cups sifted flour
⅔ cup sifted confectioners' sugar
½ teaspoon almond extract
4 ounces almonds, peeled, ground

Cut the butter into the flour with a knife, and rub in with fingertips. Add sugar and almond extract. Knead the dough. Combine with almonds.

Roll out the dough as thin as you can. Place half of it on a cookie sheet. Roll until it is ¼-inch thick, leaving at least 1 inch free on all sides. Cut into small squares. Bake in 450° oven for 8 minutes until golden. Remove from sheet with a knife immediately. Repeat with the remaining dough.

YIELDS ABOUT 3 DOZEN.

· ·

Polish Crullers

Of all the great variety of Polish pastries, crullers are best known in America and very much liked by the children and grandchildren of the old immigrants. It is hard to imagine a Polish-American party without these crisp fried cookies. In Poland they are traditionally served between Christmas and Lent, the time of the year called Carnival, when most elegant dances and parties take place.

There are several variations of the recipe. The following one comes from the city of Krakow. It is the best one I ever tried. It was given to me by an excellent cook and a gracious hostess—my friend who grew up in this ancient city.

Crullers, or *chrust,* are not difficult to make. With some experience, it will take you one hour from start to finish. The dough must be soft, elastic, and easy to roll out. It is most important to roll it out as thin as possible, using very little flour on the pastry board, and also to fry quickly.

You can make your crullers a day ahead of your party.

6 large egg yolks
1 tablespoon sugar
Dash of salt
1 tablespoon rum
2 tablespoons sour cream
1 tablespoon lemon juice
1 teaspoon lemon rind, grated
½ teaspoon vanilla
1½ cups sifted flour
½ teaspoon baking powder
1 pound shortening
1 cup confectioners' sugar

Beat the egg yolks with sugar and salt for 5 minutes. Add rum, sour cream, lemon juice, rind, and vanilla. Mix well. Add flour sifted with baking powder. Knead the dough for 10 minutes. Divide into 3 portions. Roll each portion paper thin. Cut out 1¼" × 5" strips. Cut a 2-inch-long hole in the middle of each strip, then pass one end of the strip through it. Heat the shortening in a large pan to 380°. Fry 5 strips at a time on both sides until golden. Place on tissue paper to cool. Pile up on a serving platter, sprinkle each layer generously with confectioners' sugar through a sieve.

YIELDS ABOUT 4 DOZEN.

Norwegian Crullers

It is always such a surprise to find out how many similarities there are in the culinary traditions of various nations. And each claims to have invented the best recipes, not realizing that across the border or even the sea, a neighbor may have something almost identical. Norwegian cooks, so far to the north of Poland, have a close copy of the famous Polish fried cookie. Let's try it and decide which one is better.

Make your crullers in the morning or a day ahead of the party. Keep uncovered in a cool, dry place.

6 *egg yolks*
4 *egg whites*
6 *tablespoons sugar*
6 *tablespoons whipping cream*
1 *teaspoon cardamom*
3 *to 4 cups flour*
1 *cup confectioners' sugar*

Beat the egg yolks and the egg whites with sugar for 15 minutes. Stir in with a spatula the cream mixed with cardamom. Add enough flour to make a dough soft enough to be rolled very thinly.

Roll out on a floured board. Cut into rectangles 1¼" × 3" inches. Cut a 1½-inch-long hole in the middle of each strip, then pass one end of the strip through it, and stretch the dough as much as you can.

Fry in deep fat, 350°, until golden brown on both sides. Place on tissue paper to cool. Sprinkle with confectioners' sugar through a sieve.

YIELDS ABOUT 6 DOZEN.

· ·

Greek Baklava

Baklava is for people who know where to buy filo, the paper-thin pastry sheets sold in Greek shops. In every larger American city there is at least one such store. Outside of Greek desserts, filo may be used for many things. It is excellent as bottom and top crust for cheese cakes, it makes a very good strudel. It even successfully takes place of the famous French puff pastry which is really quite similar.

Once you have learned to use filo, if you remember to work with it fast in a cool kitchen, in no time you can produce various Continental delicacies out of these handy pastry sheets.

SYRUP
3 cups sugar
3 cups water
1 ½ tablespoons lemon juice
1 ½ cups honey
1 teaspoon cinnamon
½ teaspoon cloves

BAKLAVA
1 ¾ pounds walnuts, ground
2 teaspoons cinnamon
1 pound filo
1 pound sweet butter, melted

TO MAKE SYRUP: Combine all the ingredients. Boil at low heat for 15 minutes, stirring constantly. Let cool for at least 4 hours.

TO MAKE BAKLAVA: Line buttered baking pan with one sheet of filo, brush with butter. Repeat with 5 pastry sheets. Sprinkle with nuts mixed with cinnamon. Repeat, brushing each sheet with butter and sprinkling each second sheet with nuts. Reserve 6 sheets of filo for the top layer. Brush each one and the top generously with butter. Cut into 2-inch diamond-shaped pieces.
Bake in a 350° oven for 1 hour. Pour cold syrup over hot baklava. Let stand for 5 hours before serving.

YIELDS 5 DOZEN.

Swedish Ginger Hearts

It is customary in Sweden to hang lovely cookies on Christmas trees. They are cut into different shapes—dolls, animals, trees, or hearts—and decorated with silvery icing.

DOUGH
¼ pound margarine
1 egg
⅓ cup and 1 tablespoon sugar
¾ cup syrup
1 teaspoon cinnamon
½ teaspoon cloves
1 teaspoon ginger
1 teaspoon grated orange rind
¾ teaspoon baking soda
3 cups sifted flour

ICING
1 cup confectioners' sugar
2 tablespoons water
A few drops peppermint extract
Red, green, and yellow food coloring

Beat the margarine until creamy. Whip the egg with sugar; add to the margarine by spoonfuls, still beating.
Bring the syrup with the spices to a boil; cool slightly. Combine with the margarine mixture. Add flour sifted with baking soda. Mix with a wooden spoon. Refrigerate overnight. Roll out, cut into heart shapes. Bake in 400° oven for 10 minutes or until golden brown. Combine the ingredients of the icing. Decorate ginger hearts with icing in different colors.

YIELDS ABOUT 4 DOZEN.

Christmas Stars from Finland

Finnish mothers bake many cookies for Christmas just as we do here. They don't spend so much time, however, shopping for gifts. Christmas is still an old-fashioned holiday in Finland, and hand-made gifts are the most appreciated because it is considered that it is the thought and the effort that count, not the price.

Many social gatherings take place on the second day of Christmas, and a good supply of Christmas cookies is always very handy.

3 cups sifted flour
1 teaspoon baking powder
½ pound and 2 tablespoons margarine
1 ½ cups whipping cream
1 jar prune pastry filling
1 egg, lightly beaten
½ cup sugar

Sift the flour with baking powder. Cut the margarine into the flour with a knife. Add whipped cream. Knead the dough quickly. Cover and refrigerate for ½ hour.

Roll out on a floured board ⅛-inch thick. Cut into 2½-inch squares. Place a teaspoon of prune filling in the center of each square. Split the corners, and fold over every other strip so that they meet in the center. Brush with egg, sprinkle with sugar, and bake in 425° oven for about 13 minutes until delicately browned.

YIELDS ABOUT 4 DOZEN.

. . . .

Almond Crescents from Rumania

Young girls love to bake cookies. This is a simple Rumanian recipe that even a 12-year-old can easily follow.

2 cups flour, sifted
½ cups confectioners' sugar, sifted
½ pound sweet butter, soft
1 egg yolk
1 teaspoon vanilla extract
½ cup peeled almonds, ground
¾ cup confectioners' sugar

Measure all the ingredients. Combine the flour with sugar. Cut in the butter with a knife. Add the egg yolk and vanilla. Mix well. Add almonds. Knead the dough. Refrigerate for 1 hour.

Make a roll. Cut small slices and form crescents. Bake in 350° oven for 25 minutes. Roll immediately in sugar.

YIELDS ABOUT 3 DOZEN.

. . . .

Walnut Cookies from Prague

It used to be a tedious job to pour cookies into different shapes from a pastry bag. With a cookie gun, life can be more pleasant, and we can still create and enjoy the old-fashioned goodies.

2 egg whites, lightly beaten
2 cups confectioners' sugar
2 cups walnuts, ground
½ cup bread crumbs
1 teaspoon rum

Combine all the ingredients. Sprinkle a variety of small shapes onto a buttered cookie sheet. Bake in 275° F. oven for about 40 minutes.

YIELDS ABOUT 6 DOZEN.

Index